Konstantin Alekseevich Shapovalov

Light scattering by nonspherical particles in the RGD approximation

AF153239

Konstantin Alekseevich Shapovalov

Light scattering by nonspherical particles in the RGD approximation

Single scattering

LAP LAMBERT Academic Publishing

Impressum / Imprint
Bibliografische Information der Deutschen Nationalbibliothek: Die Deutsche Nationalbibliothek verzeichnet diese Publikation in der Deutschen Nationalbibliografie; detaillierte bibliografische Daten sind im Internet über http://dnb.d-nb.de abrufbar.
Alle in diesem Buch genannten Marken und Produktnamen unterliegen warenzeichen-, marken- oder patentrechtlichem Schutz bzw. sind Warenzeichen oder eingetragene Warenzeichen der jeweiligen Inhaber. Die Wiedergabe von Marken, Produktnamen, Gebrauchsnamen, Handelsnamen, Warenbezeichnungen u.s.w. in diesem Werk berechtigt auch ohne besondere Kennzeichnung nicht zu der Annahme, dass solche Namen im Sinne der Warenzeichen- und Markenschutzgesetzgebung als frei zu betrachten wären und daher von jedermann benutzt werden dürften.

Bibliographic information published by the Deutsche Nationalbibliothek: The Deutsche Nationalbibliothek lists this publication in the Deutsche Nationalbibliografie; detailed bibliographic data are available in the Internet at http://dnb.d-nb.de.
Any brand names and product names mentioned in this book are subject to trademark, brand or patent protection and are trademarks or registered trademarks of their respective holders. The use of brand names, product names, common names, trade names, product descriptions etc. even without a particular marking in this works is in no way to be construed to mean that such names may be regarded as unrestricted in respect of trademark and brand protection legislation and could thus be used by anyone.

Coverbild / Cover image: www.ingimage.com

Verlag / Publisher:
LAP LAMBERT Academic Publishing
ist ein Imprint der / is a trademark of
OmniScriptum GmbH & Co. KG
Heinrich-Böcking-Str. 6-8, 66121 Saarbrücken, Deutschland / Germany
Email: info@lap-publishing.com

Herstellung: siehe letzte Seite /
Printed at: see last page
ISBN: 978-3-659-61634-1

Contents

List of abbreviations

AD   Anomalous Diffraction
ADDA C implementation of the Discrete Dipole Approximation
DDA   Discrete Dipole Approximation
EBCM Extended Boundary Condition Method
GPMT Generalized Point-Matching Technique
RGD   Rayleigh-Gans-Debye

Introduction

The light scattering and absorption of electromagnetic radiation is widely used in different branches of science and engineering for the study of structure and properties of inhomogeneous media. In recent years because of great significance such applications as optics of atmosphere and ocean, radio wave propagation, radio communication, physical chemistry of solutions and colloids, biophysics, laser biomedicine the theory and practice of light scattering techniques have been sufficiently developed [1-6].

When we solve light scattering problem by aerosols particles of atmosphere and so on these particles are modeled by particles of different geometrical shape. So, for spherical particles by separation of variables analytical decision (solution) or theory Mie [1, 2] is obtained and well-known.

Cloud droplets have a spherical shape in most cases, although particles of other shapes can exist due to various external influences. However, close on spherical shape have liquid aerosols. Dust and soot particles, ice crystals of clouds have strong nonspherical shape. For example, ice crystals of cirrus clouds have cylindrical shape and model by hexagonal prisms. But even on modern step of development of analytical and calculation methods of evaluation of characteristics of light scattering ensembles and single particle of nonspherical shape are frequently encountered with serious difficulties [3].

But if a particle has other than a regular geometrical shape, then it is difficult or impossible to solve the scattering problem analytically in its most general form that oblige to use numerical and approximate analytical methods.

Therefore, if particles of dispersion media are optically "soft" $|m-1| \ll 1$, where m is a relative refractive index of light scattering particle (or particles are suspended in a medium with similar optical properties), then we can use suitable approximate methods of Rayleigh-Gans-Debye (RGD) or Anomalous Diffraction (AD) [1, 2, 4-11]. Note that the domain of validity of the RGD approximation is differed from the AD approximation [2, 4, 8, 12, 13].

There have been some attempts to apply the RGD approximation to a particle of completely arbitrary shape and size, none of these has been truly satisfactory [14, 15], because of it leads us to a numerical solution by means of Fourier transformation. The analytical expressions are preferred by reason of they have more precise results and can serve as a basis for rigorous solution [16, 17]. See also discussion about advantages of analytical approximate solutions over numerical solution in [11].

However, most of scientists consider the RGD approximation is so simple and don't devote his time for this approximation. I hope this new book is filled up the gaps about the RGD approximation.

The book consists of ten chapters. Chapter **1** contains a formulation of the light scattering problem, the main ideas of the method, and a brief description of the general approach for a composite or compound particle. Chapter **2** contains some results for a spheroid and ellipsoid in the RGD approximation. Chapters **3** and **5** contain some earlier results for a circular cylinder and cone as particles similar to prism, pyramid, respectively. Chapters **4** and **6** contain new results for a prisms and pyramids in the RGD approximation. Chapter **7** contains results for torus and toroidal particles in the RGD approximation. Chapter **8** contains new results for shell particles in the RGD approximation. Chapter **9** contains some data for aggregates of particles and for polydisperse media of particles with different sizes. Chapter **10** contains concluding remarks.

1. General approach for the obtaining form factor and light scattering amplitude of compound or multilayered particle in the RGD approximation

Consider that a particle illuminates by a plane electromagnetic wave. Use integral expression of amplitude of light scattering in the RGD approximation (or the first Born approximation) in a scalar form [8, 18]:

$$f(\theta,\beta) = \frac{k^2 |\mathbf{P}|}{4\pi} \int_V (m^2 - 1) \exp(i\,\mathbf{k_s} \cdot \mathbf{r}) dV, \tag{1}$$

where $\mathbf{i}, \mathbf{s}$ are unit vectors along directions of the incident and scattering light respectively, $\mathbf{r}$ is the radius-vector of a point inside the particle, $\mathbf{k_s} = k(\mathbf{i} - \mathbf{s})$, $k = 2\pi/\lambda$ is the wave number, λ is the wavelength of light, $|\mathbf{k_s}| = 2k\sin\left(\frac{\theta}{2}\right)$, θ is the angle between vectors $\mathbf{i}$ and $\mathbf{s}$, β is the angle between axis z and vector $\mathbf{k_s}$, $|\mathbf{P}| = |[-\mathbf{s} \times (\mathbf{s} \times \mathbf{e_i})]|$, $\mathbf{e_i}$ is the unit vector along direction of the incident light polarization, (forth for brief text in a scalar form $|\mathbf{P}| = 1$).

Note that the amplitude can be expressed another way in terms of the angles in spherical coordinates pointed direction of the incident θ_i, ϕ_i and scattering light θ_s, ϕ_s separately: $k_1 = k(\sin\theta_i \cos\phi_i - \sin\theta_s \cos\phi_s)$,

$k_2 = k(\sin\theta_i \sin\phi_i - \sin\theta_s \sin\phi_s)$, $k_3 = k(\cos\theta_i - \cos\theta_s)$, $k_4 = \sqrt{k_1^2 + k_2^2}$,

$k_s = \sqrt{k_1^2 + k_2^2 + k_3^2}$.

The form factor in the RGD approximation [1, 2, 4] for a homogeneous particle with the volume V may be written as

$$\Phi(\theta,\beta) = \frac{4\pi\, f(\theta,\beta)}{k^2(m^2-1)V} = \frac{1}{V} \int_V \exp(i\,\mathbf{k_s} \cdot \mathbf{r}) dV. \tag{2}$$

Before we give some properties of light scattering amplitude note that the RGD approximation is valid when so-called "phase shift" of central ray Δ is much smaller compared with unity ($\Delta = 2ka\,|m\text{-}1|<<1$, where a is the longest dimension through the particle) [1, 2, 4, 8, 18].

Firstly, for a composite particle, containing q layers or distinct nonoverlapping regions [1, 2], we get

$$f(\theta,\beta) = \frac{k^2}{4\pi}\left[(m_q^2 - 1)V_q\Phi_q(\theta,\beta) + \sum_{j=1}^{q-1}(m_j^2 - m_{j+1}^2)V_j\Phi_j(\theta,\beta) \right], \tag{3}$$

5

where j is the number of layer (or region), m_j is the relative refractive index of the j th layer, V_j is its volume, $\Phi_j(\theta,\beta)$ is the form factor of the j th layer.

Secondly, if a particle with the form factor $\Phi_0(\theta,\beta)$ shifts from center of coordinates to the position pointed by a vector $\mathbf{r_M}$, then we can obtain the form factor as a multiplication by $\exp(i\,\mathbf{k_s}\cdot\mathbf{r_M})$ [19, 20]:

$$\Phi_M(\theta,\beta) = \frac{1}{V}\int_V \exp(i\,\mathbf{k_s}\cdot(\mathbf{r}+\mathbf{r_M}))dV = \exp(i\,\mathbf{k_s}\cdot\mathbf{r_M})\Phi_0(\theta,\beta). \qquad (4)$$

Thirdly, if a particle rotates, for example, about axis OZ on Eulerian angle γ and because Jacobian of transformation for such rotation in Eq. (2) is equal to 1, then we can obtain the form factor by transforming only expressions of k_1, k_2 into new position $k_1(\gamma)$, $k_2(\gamma)$ as follows:

$$\begin{pmatrix} k_1(\gamma) \\ k_2(\gamma) \end{pmatrix} = \begin{pmatrix} \cos\gamma & -\sin\gamma \\ \sin\gamma & \cos\gamma \end{pmatrix}\begin{pmatrix} k_1 \\ k_2 \end{pmatrix}. \qquad (5)$$

Thus, we establish rotational-translational properties of light scattering amplitude (not available in literature for the RGD approximation): shift or translation in Eq. (4), rotation (see Eq. (5)). Eqs. (3)-(5) provide us a convenient way to construct form factor in the RGD approximation for a compound particle and for a system of particles if the form factors of every particle are known.

2. Spheroid and Ellipsoid

The equations for the calculation of light scattering characteristics: amplitude, phase function and others for spheroid and ellipsoid in the RGD approximation (or in first Born approximation) are well-known and used earlier by different authors [2, 4, 8].

For a homogeneous ellipsoid with half axis a, b, c (see Fig. 1 a) the amplitude of light scattering [4, 8] is a well-known and may be written as

$$f_{ELL} = \frac{k^2}{2\pi}(m^2 - 1) \cdot V_{ELL} \cdot \frac{3 j_1\left(\sqrt{(k_1 a)^2 + (k_2 b)^2 + (k_3 c)^2}\right)}{\sqrt{(k_1 a)^2 + (k_2 b)^2 + (k_3 c)^2}}, \tag{6}$$

where $V_{ELL} = \frac{4}{3}\pi abc$ is the volume of ellipsoid, $j_1(x) = \dfrac{\sin x - x \cos x}{x^2}$ is a spherical Bessel function of first order.

Also for a homogeneous spheroid with axis of rotation along OZ c and radius $R = a = b$ (see Fig. 1 b) the amplitude of light scattering [4, 8] follows from Eq. (6):

$$f_{SPH} = \frac{k^2}{2\pi}(m^2 - 1) \cdot V_{SPH} \cdot \frac{3 j_1\left(\sqrt{(k_4 R)^2 + (k_3 c)^2}\right)}{\sqrt{(k_4 R)^2 + (k_3 c)^2}}, \tag{7}$$

where $V_{SPH} = \frac{4}{3}\pi R^2 c$ is the volume of spheroid.

So, the phase function [or element of scattering matrix f_{11}] for natural incident light (unpolarized or arbitrary polarized light) is calculated by a formula [1,2,8]

$$f_{11}(\theta) = \frac{1 + \cos^2(\theta)}{2} k^2 |f(\theta)|^2, \tag{8}$$

where $|f(\theta)|^2$ is a square of modulus of light scattering amplitude.

Further, the light scattering phase function is normalized on value in forward direction. Phase functions calculated for incident light perpendicular and along axis of spheroid's symmetry for oblate and prolate spheroids in the RGD approximation and in the rigorous solutions by technique of T-matrix [3, 21] and ADDA [22] (ADDA is a C implementation of the discrete dipole approximation (DDA) [23]) with size kR=2 and kc=0.2, kc=6 are shown in Figs. 2, 3. Note, the rigorous solution by technique of T-matrix, developed in [3, 21], is modified method of extended boundary condition method (EBCM).

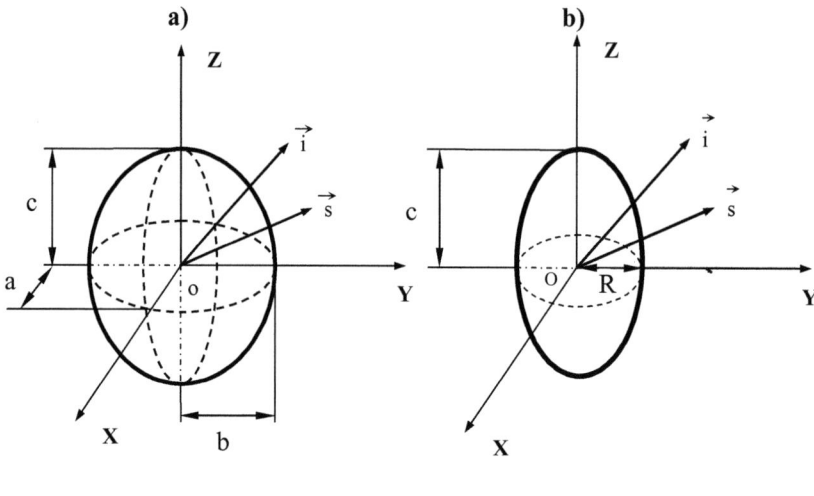

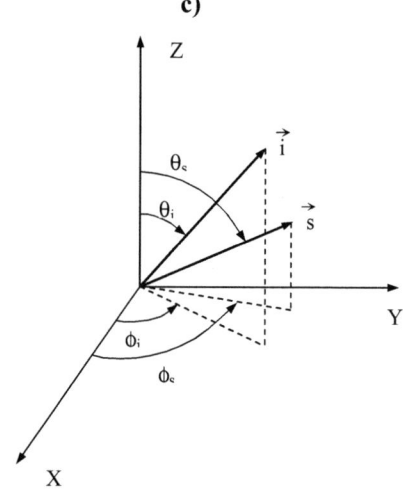

Fig. 1. Geometry of light scattering by ellipsoid (a) and spheroid (b) and coordinate system (c).

A)

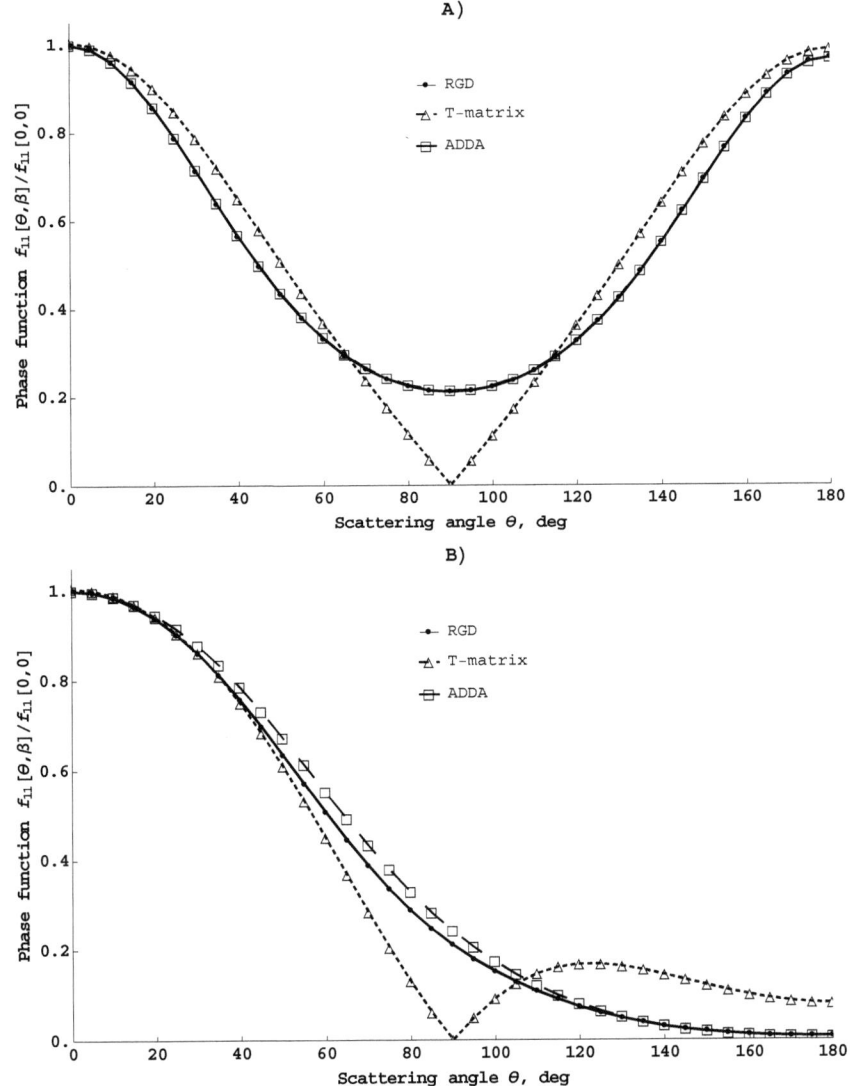

Fig. 2. Normalized phase function $f_{11}(\theta)/f_{11}(0)$ vs. scattering angle θ for an oblate spheroid in the RGD approximation, T-matrix and ADDA solution of spheroid with relative refractive index $m=1.1+i\cdot0.01$ provided $kR=2$, $kc=0.2$ and for incident light along (A) perpendicular (B) to axis of spheroid's symmetry.

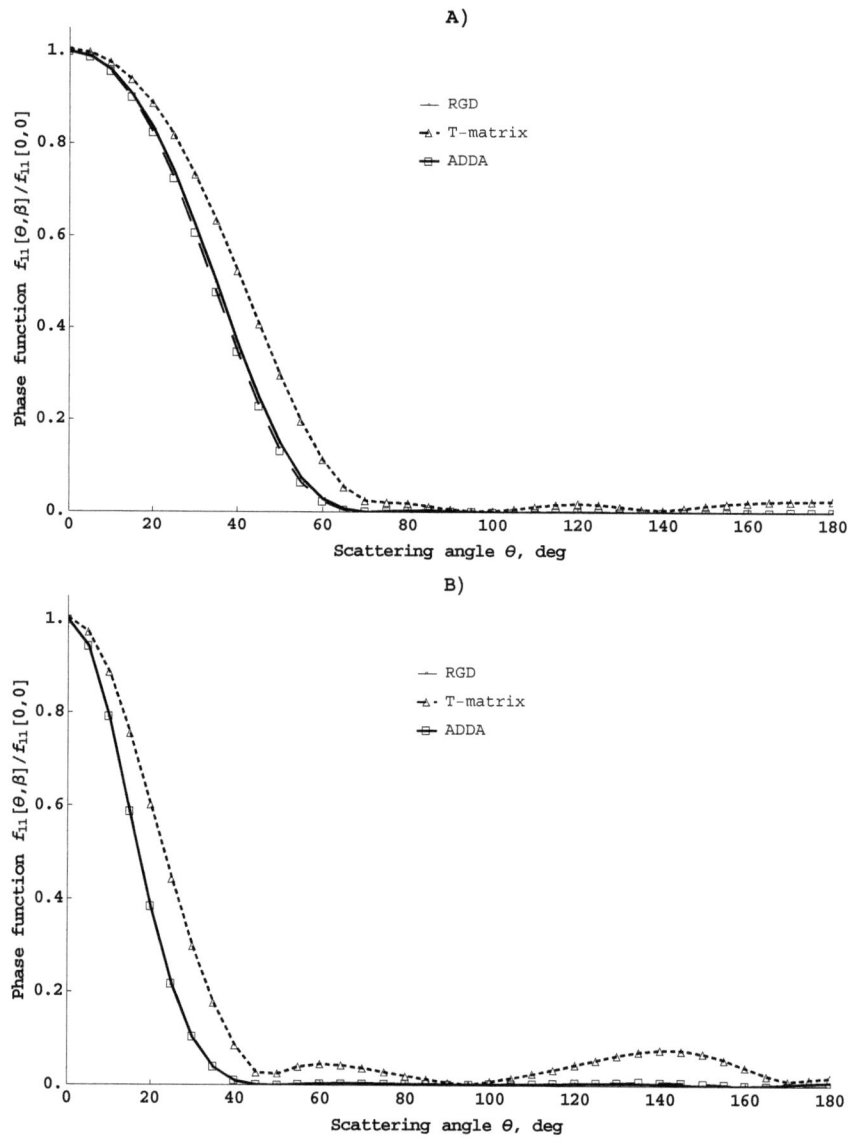

Fig. 3. Normalized phase function $f_{11}(\theta)/f_{11}(0)$ vs. scattering angle θ for a prolate spheroid in the RGD approximation, T-matrix and ADDA solution of spheroid with relative refractive index m=1.1+$i\cdot$0.01 provided kR =2, kc =6 and for incident light along (A) perpendicular (B) to axis of spheroid's symmetry.

3. Circular cylinder

The formulas for the calculation of light scattering characteristics: amplitude, phase function and others for a cylinder in the RGD approximation (or in first Born approximation) are obtained earlier [1, 4, 24, 25].

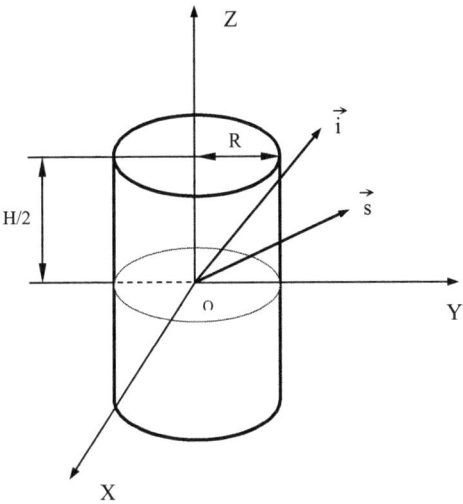

Fig. 4. Geometry of light scattering by a cylinder [24].

For a homogeneous cylinder with height H and radius R (see Fig. 4) the amplitude of light scattering [1, 4, 18, 24] is a well-known and may be written as

$$f_{CYL} = \frac{k^2}{2\pi}(m^2 - 1) \cdot V_{CYL} \cdot j_0\left(\frac{k_3 H}{2}\right) \frac{J_1(k_4 R)}{k_4 R},$$ (9)

where $V_{CYL} = \pi R^2 H$ is the volume of circular cylinder, $J_1(x)$ is a Bessel function of first order, $j_0(x) = \dfrac{\sin x}{x}$ is a spherical Bessel function of zero order.

Phase functions calculated for incident light perpendicular and along axis of cylinder's symmetry in the RGD approximation and for cylinders in the rigorous solutions by method of T-matrix [3,21], ADDA [22] and separation of variables for infinitely long cylinder [1] are shown in Figs. 5,6.

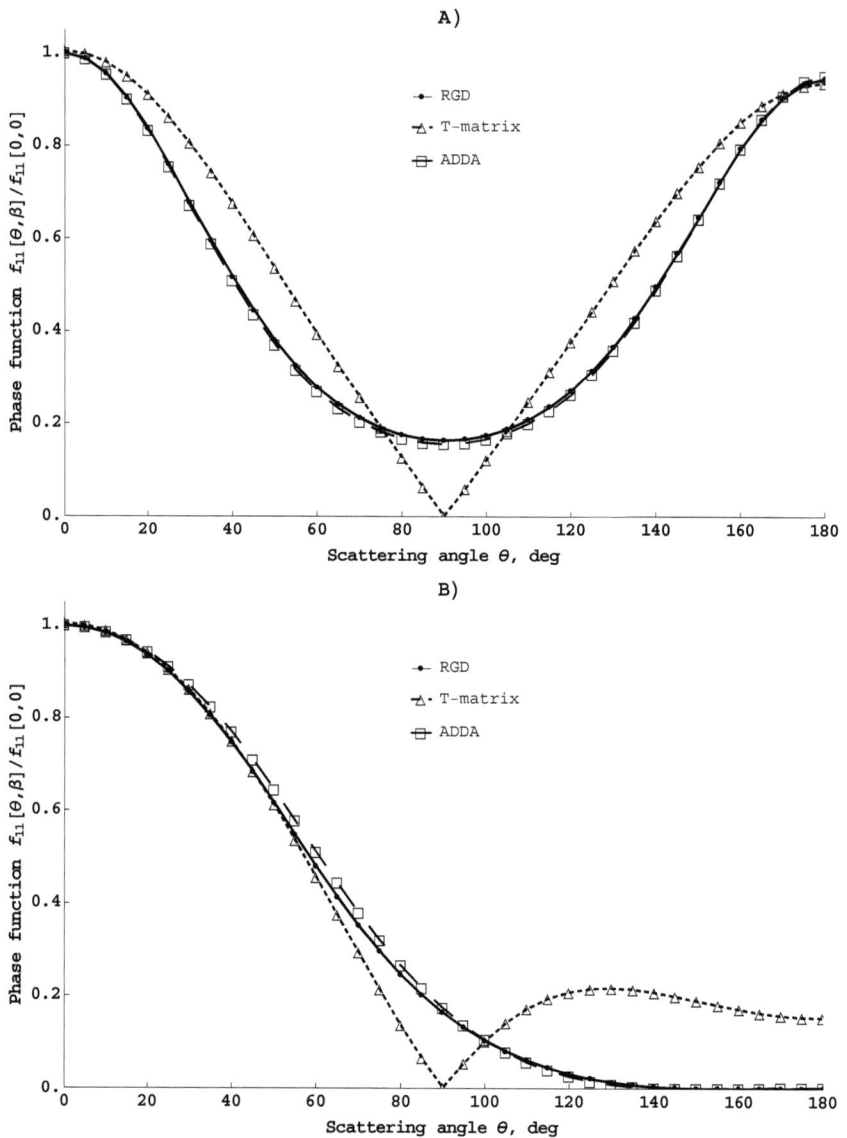

Fig. 5. Normalized phase function $f_{11}(\theta)/f_{11}(0)$ vs. scattering angle θ for a cylinder in the RGD approximation, T-matrix and ADDA solution of cylinder with relative refractive index m=1.1+i·0.01 provided kR =2, kH =0.4 and for incident light along (A) perpendicular (B) to axis of cylinder's symmetry.

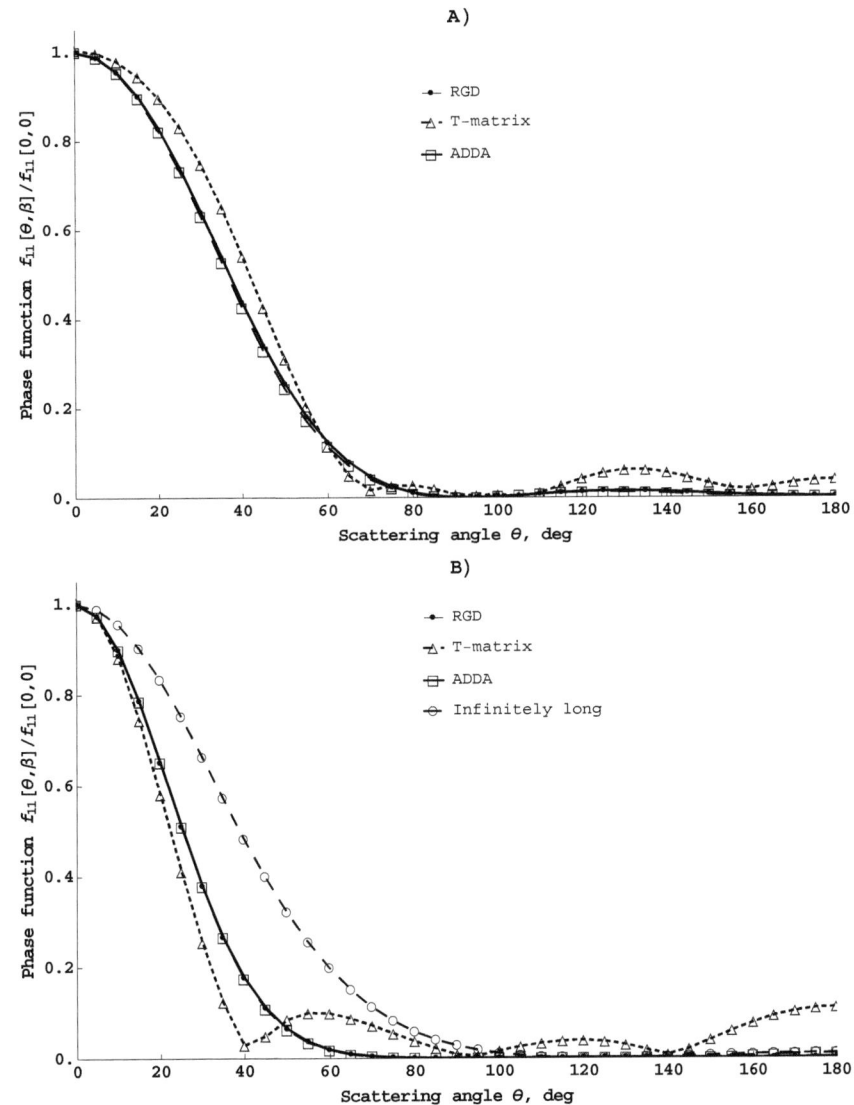

Fig. 6. Normalized phase function $f_{11}(\theta)/f_{11}(0)$ vs. scattering angle θ for a cylinder in the RGD approximation, T-matrix, ADDA and for infinitely long cylinder [1] solution with relative refractive index m=1.1+i·0.01 provided   kR =2, kH =6 and for incident light along (A) perpendicular (B) to axis of cylinder's symmetry.

Light scattering cross section σ_s [1, 8], normalized to cross section area of projection S of particle on a plane, which is perpendicular to the axis of beam, (or light scattering efficiency factor Q_s) is equal

$$\frac{\sigma_s}{S} = Q_s = \frac{\int\limits_{4\pi} |f(s,i)|^2 d\omega}{S},$$ (10)

where $d\omega$ is an element of solid angle (in spherical system of coordinates we can write $\sin(\theta_s)d\theta_s d\phi_s$).

Thus, the area of projection S of particle on a plane, which is perpendicular to the axis of beam, for circular cylinder [25] is equal

$$S(\theta_i,\phi_i)=2\,R\cdot[\,H\sin(\theta_i)+\frac{\pi}{2}R\cdot\cos(\theta_i)]\,.$$ (11)

Light scattering efficiency factors Q_s calculated for incident light perpendicular and along axis of cylinder's symmetry in the RGD approximation and ADDA with relative refractive index $m=1.31+i\cdot0.01$ are shown in Fig. 7.

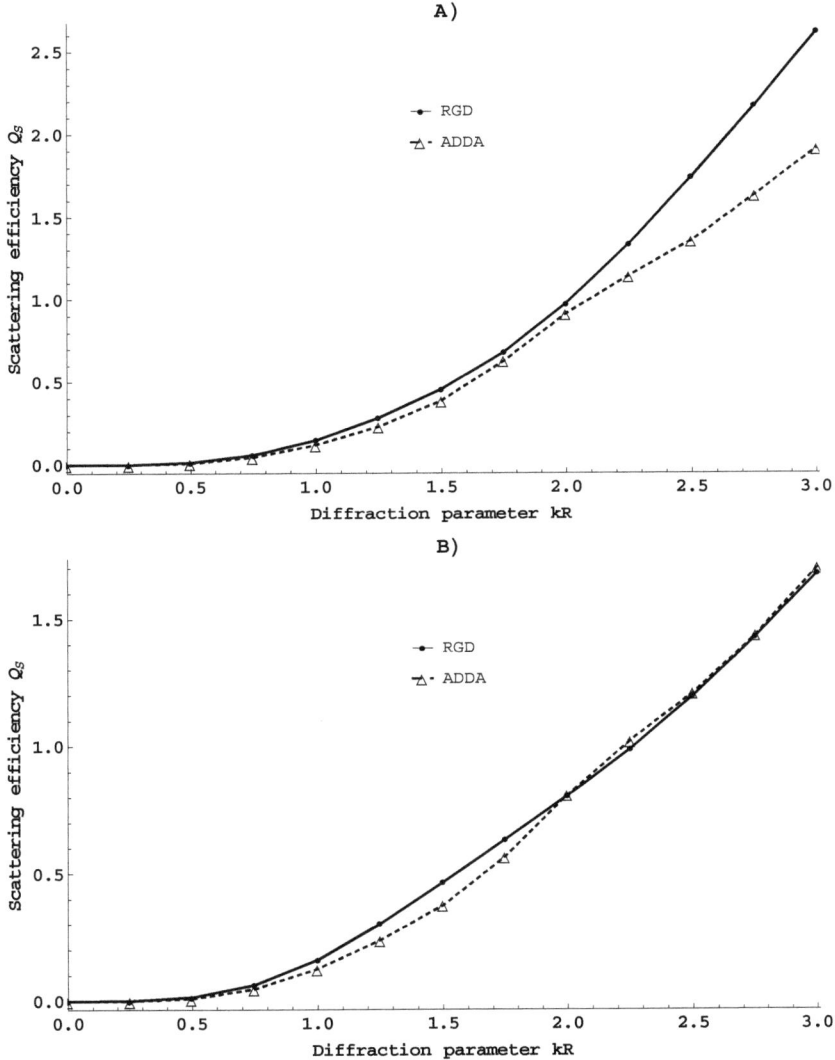

Fig. 7. Light scattering efficiency Q_S vs. diffraction parameter kR in the RGD approximation and ADDA for circular cylinders with relative refractive index m=1.31+i·0.01 provided aspect ratio H/(2R) = 1 and direction of incident light along axis of symmetry β =0 (A) and perpendicular β =π/2 (B).

4. Polygonal prism and hexagonal cylinder

Analytic equations for the light scattering amplitude in the RGD approximation of a prism (column) consisting of an arbitrary polygonal base may be obtained using Eqs. (3),(5) [19, 20].

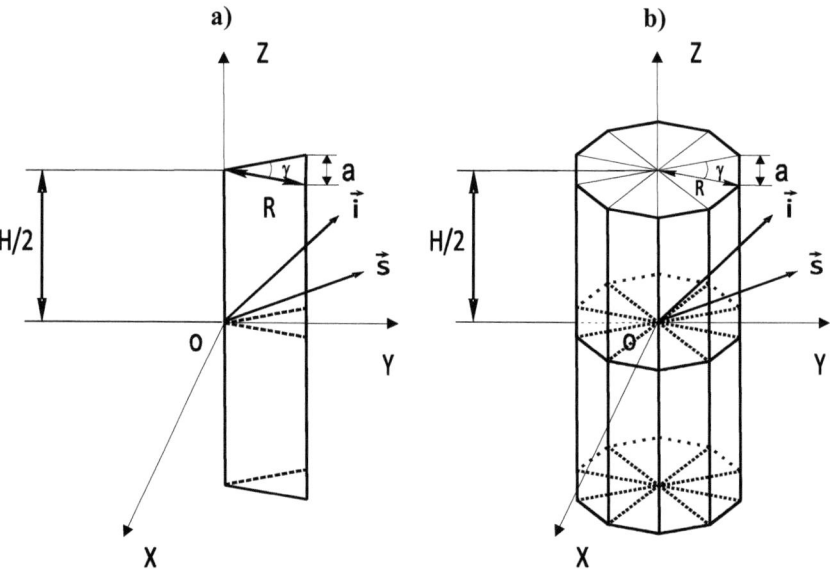

Fig. 8. Geometry of light scattering by a wedge of polygonal prism (a) and whole polygonal prism (b), consisting of such wedges (n=10) [19].

First of all, it's necessary to obtain expression of light scattering amplitude for elementary polygon segment wedge (see Fig. 8 a)), having angle γ, radius of circumscribed circle R and side a, with full height H. Then, rotating wedges on angle γ-fold and summing amplitudes of wedges in new position, we can get general light scattering amplitude for whole prism (see Fig. 8 b)).

So, amplitude for polygonal prism segment wedge [19] is equal

$$f_{PM} = \frac{k^2(m^2-1)}{4\pi}\frac{V_{PM}}{k_1 R \sin\gamma'} j_0\!\left(k_3\frac{H}{2}\right)\![h_0(k_5 R)-h_0(k_6 R)+i\,(j_0(k_6 R)-j_0(k_5 R))], \qquad (12)$$

where $V_{PM} = \dfrac{1}{2}HR^2\sin\gamma$ is a volume of prism wedge, $\gamma = \dfrac{2\pi}{n}, \gamma' = \dfrac{\gamma}{2}$,

$k_5 = k_2\cos\gamma' + k_1\sin\gamma'$, $\quad k_6 = k_2\cos\gamma' - k_1\sin\gamma'$, $\quad R$ is a radius of circumscribed circle, n is a number of segment of polygon, $j_0(x)$, $h_0(x)$ are spherical Bessel and Struve functions of zero order.

Rotating about axis OZ $n-1$ times light scattering amplitude for wedge prism (12) using Eq. (5) and summing all terms, we obtain

$$f_{PRISM} = \sum_{s=0}^{n-1} f_{PM}(s\,\gamma). \qquad (13)$$

For hexagonal cylinder (see parameters Fig. 9) we easy obtain [24]:

$$f_{HEX} = \frac{k^2(m^2-1)}{6\pi} \cdot V_{HEX} \cdot j_0\left(\frac{k_3 H}{2}\right)[F_1 + F_2 + F_3], \qquad (14)$$

where $V_{HEX} = \dfrac{3 \cdot \sqrt{3}}{2} R^2 H$ is the volume of hexagonal cylinder,

$$F_1 = j_0\left(\frac{k_1 R}{2}\right)j_0\left(\frac{k_2\sqrt{3}R}{2}\right), F_2 = \frac{1}{4}\left(1-\sqrt{3}\frac{k_1}{k_2}\right)j_0\left(\frac{\sqrt{3}R(k_2-\sqrt{3}k_1)}{4}\right)j_0\left(\frac{R(k_1+\sqrt{3}k_2)}{4}\right),$$

$$F_3 = \frac{1}{4}\left(1+\sqrt{3}\frac{k_1}{k_2}\right)j_0\left(\frac{\sqrt{3}R(k_2+\sqrt{3}k_1)}{4}\right)j_0\left(\frac{R(k_1-\sqrt{3}k_2)}{4}\right).$$

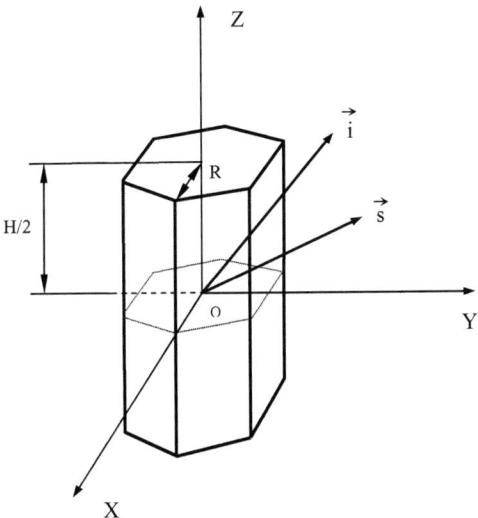

Fig. 9. Geometry of light scattering by a hexagonal prism [24].

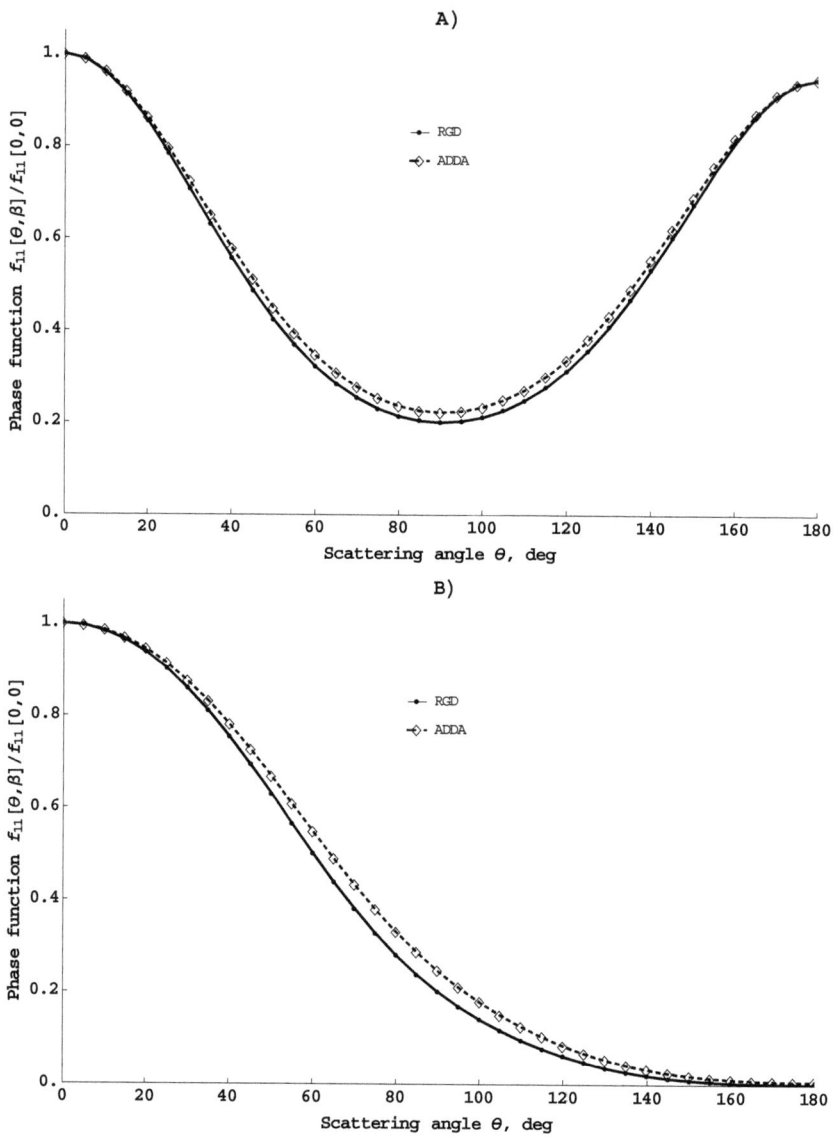

Fig. 10. Normalized phase function $f_{11}(\theta)/f_{11}(0)$ vs. scattering angle θ for a hexagonal cylinder in the RGD approximation and ADDA solution with relative refractive index m=1.1+i·0.01 provided kR =2, kH =0.4 and for incident light along (A) and perpendicular (B) to axis of cylinder's symmetry.

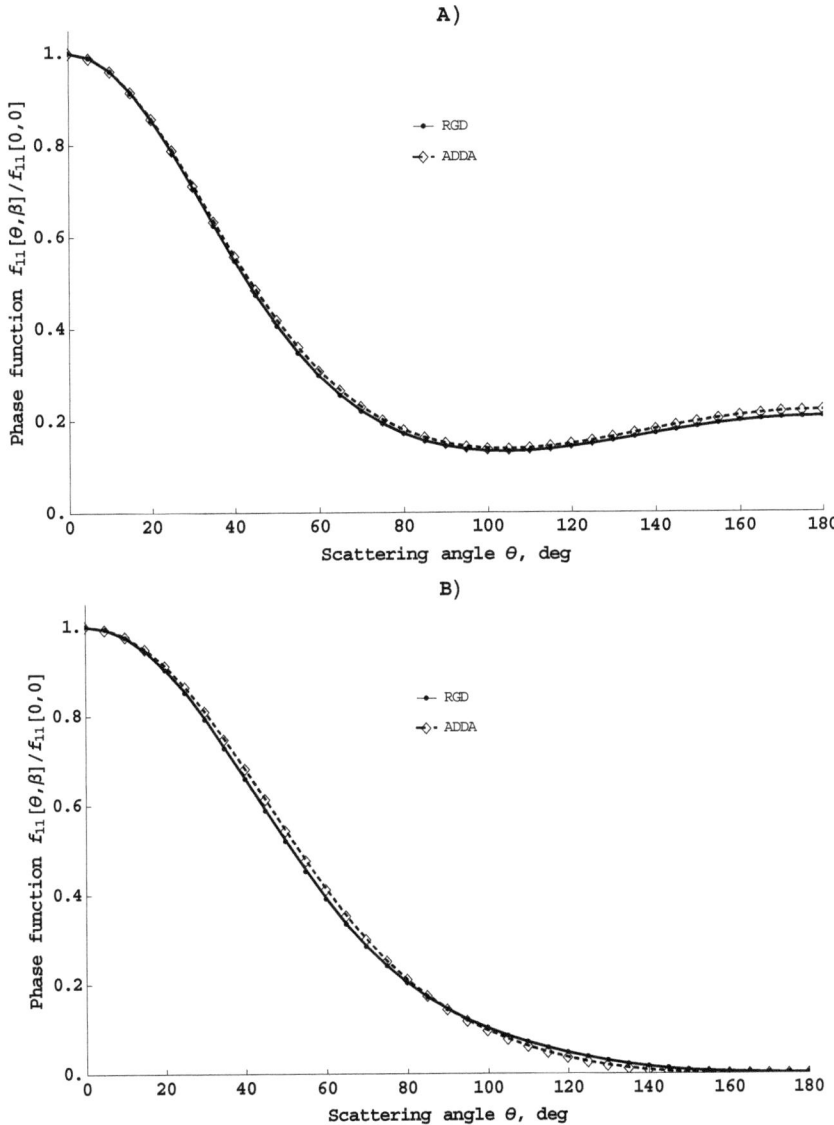

Fig. 11. Normalized phase function $f_{11}(\theta)/f_{11}(0)$ vs. scattering angle θ for a hexagonal cylinder in the RGD approximation and ADDA solution with relative refractive index $m=1.1+i\cdot0.01$ provided kR =2, kH =2 and for incident light along (A) and perpendicular (B) to axis of cylinder's symmetry.

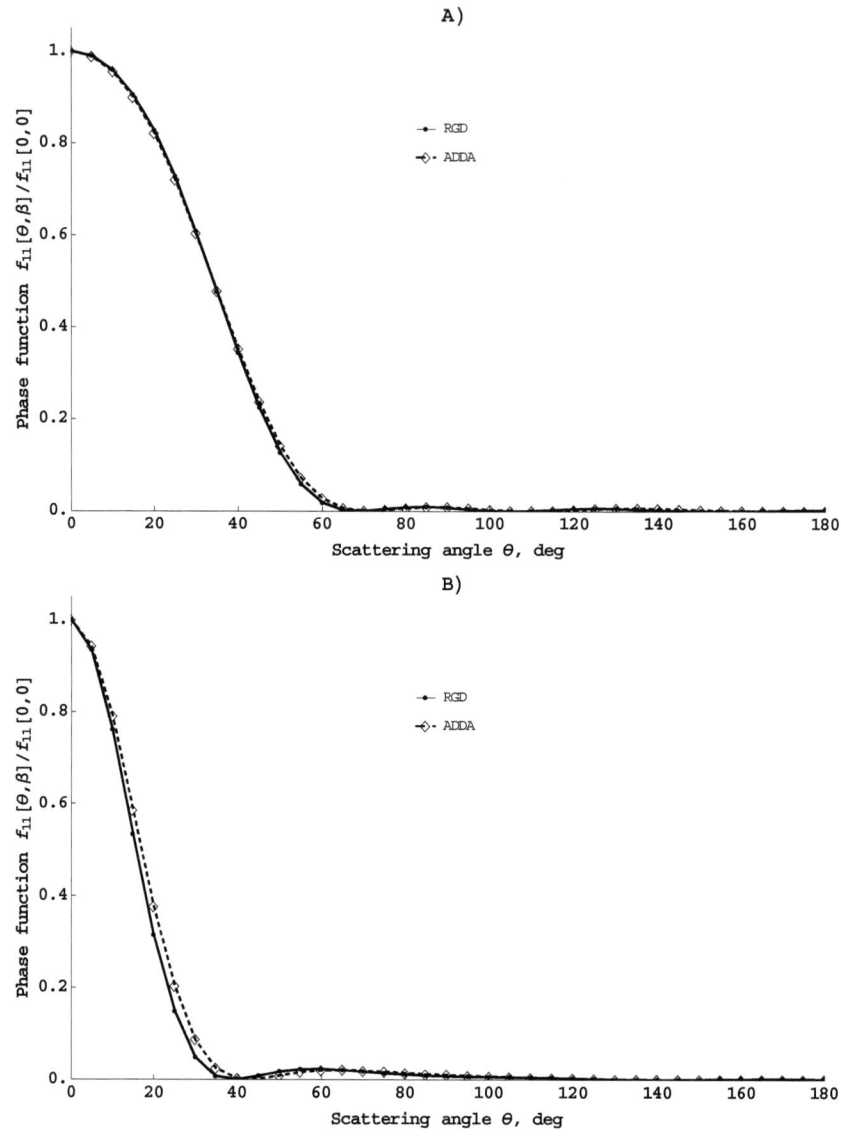

Fig. 12. Normalized phase function $f_{11}(\theta)/f_{11}(0)$ vs. scattering angle θ for a hexagonal cylinder in the RGD approximation and ADDA solution with relative refractive index m=1.1+i·0.01 provided kR =2, kH =10 and for incident light along (A) and perpendicular (B) to axis of cylinder's symmetry.

Phase functions calculated for incident light perpendicular and along axis of cylinder's symmetry in the RGD approximation and ADDA under size kR=2 and with relative refractive index m=1.1+i·0.01 are shown in Figs. 10,11,12.

The area of projection S of particle on a plane, which is perpendicular to the axis of beam, for hexagonal cylinder [25] is equal

$$S(\theta_i,\phi_i)=2 \, R \cdot \cos\left(\phi_i - \frac{\pi}{6}(2p-1)\right) \cdot [H\sin(\theta_i)+\frac{3}{2}R \cdot \cos(\theta_i)], \tag{15}$$

where p is a number from 1 to 6, depending upon interval of changed angle ϕ_i (so within $0 \leq \phi_i < \frac{\pi}{3}$ p=1, within $\frac{\pi}{3} \leq \phi_i < \frac{2\pi}{3}$ p=2 and etc).

Light scattering efficiency factors Q_s calculated for incident light perpendicular and along axis of hexagonal cylinder's symmetry in the RGD approximation and ADDA with relative refractive index m=1.31+i·0.01 are shown in Fig. 13. Some discrepancy of light scattering efficiency factors Q_s (Fig. 13) in the RGD and ADDA is existed when increment of the diffraction parameter kR.

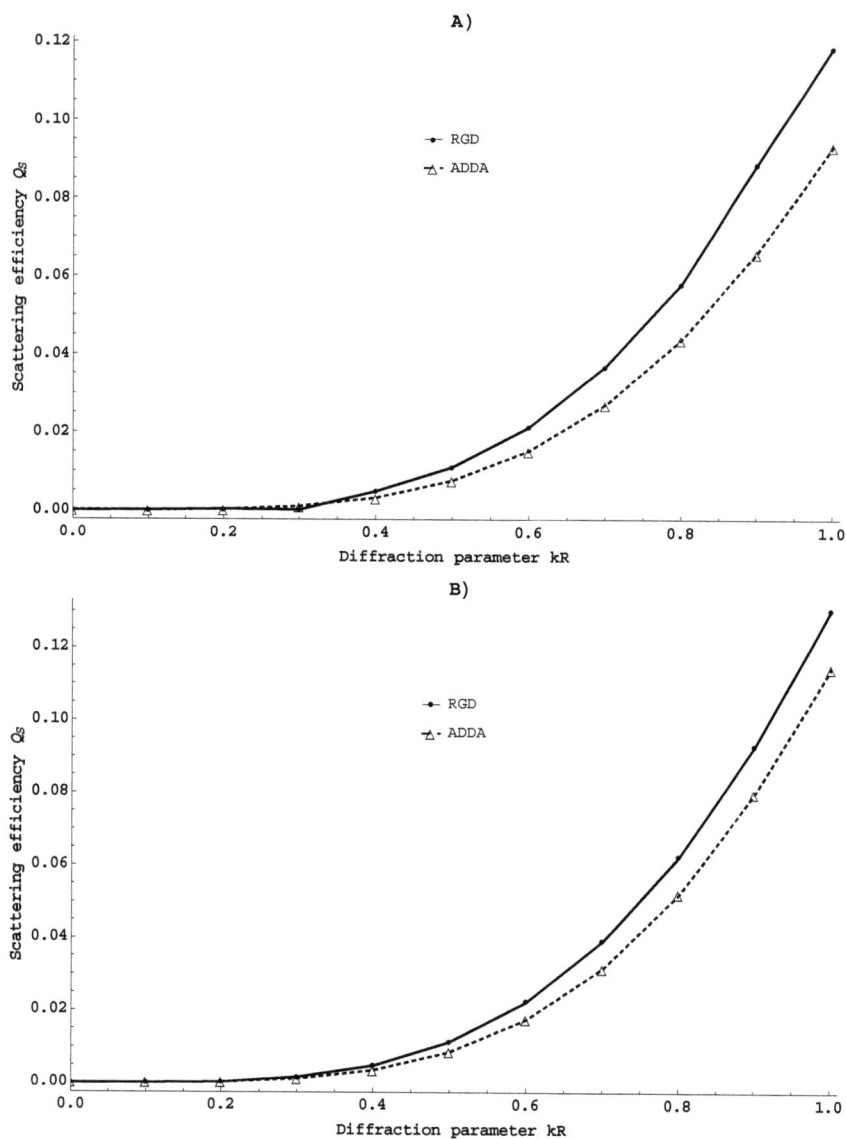

Fig. 13. Light scattering efficiency Q_S vs. diffraction parameter kR in the RGD approximation and ADDA for hexagonal cylinder with relative refractive index $m=1.31+i\cdot0.01$ provided aspect ratio $H/(2R) = 1$ and direction of incident light along axis of symmetry $\beta = 0$ (A) and perpendicular $\beta = \pi/2$ (B).

5. Cone

The formulas for the calculation of light scattering amplitude for a cone in the RGD approximation are obtained earlier in [19, 20, 26].

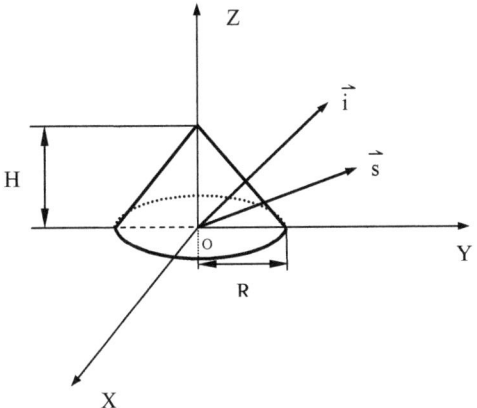

Fig. 14. Geometry of light scattering by a cone [19].

For case $k_4=0$, (in general case only expanded in series is available) [20]:

$$f_{CONE} = \frac{k^2(m^2-1)3V_{CONE}}{2\pi k_3 H}\left[h_0(k_3 H) - j_1(k_3 H) + i\left(1 - h_1(k_3 H) - j_0(k_3 H)\right)\right], \quad (16)$$

where $j_0(x)$, $j_1(x)$, $h_0(x) = \dfrac{1-\cos x}{x}$, $h_1(x) = \dfrac{1}{2} + \dfrac{1-\cos x - x\sin x}{x^2}$ are spherical Bessel and Struve functions of zero and first order, $k_3 = k_S \cos\beta$, $k_4 = k_S \sin\beta$, $V_{CONE} = \pi R^2 H / 3$.

Furthermore, for special case $k_3 H = k_4 R$ [20] light scattering amplitude in RGD is equal

$$f_{CONE} = \frac{k^2(m^2-1)V_{CONE}}{2\pi}\frac{\exp(ik_3 H)}{k_3 H}[f_1 + if_2], \quad (17)$$

where $f_1 = \cos(k_3 H)J_1(k_3 H) + \sin(k_3 H)J_2(k_3 H)$,
$f_2 = \cos(k_3 H)J_2(k_3 H) - \sin(k_3 H)J_1(k_3 H)$.

And light scattering phase functions calculated using Eq. (17) for particles with relative refractive index $m=1.1+i\cdot0.01$ are shown in Fig. 15. In contrast to [19] curves for cone are corrected because this paper has an error in denominator in Eq. (17).

23

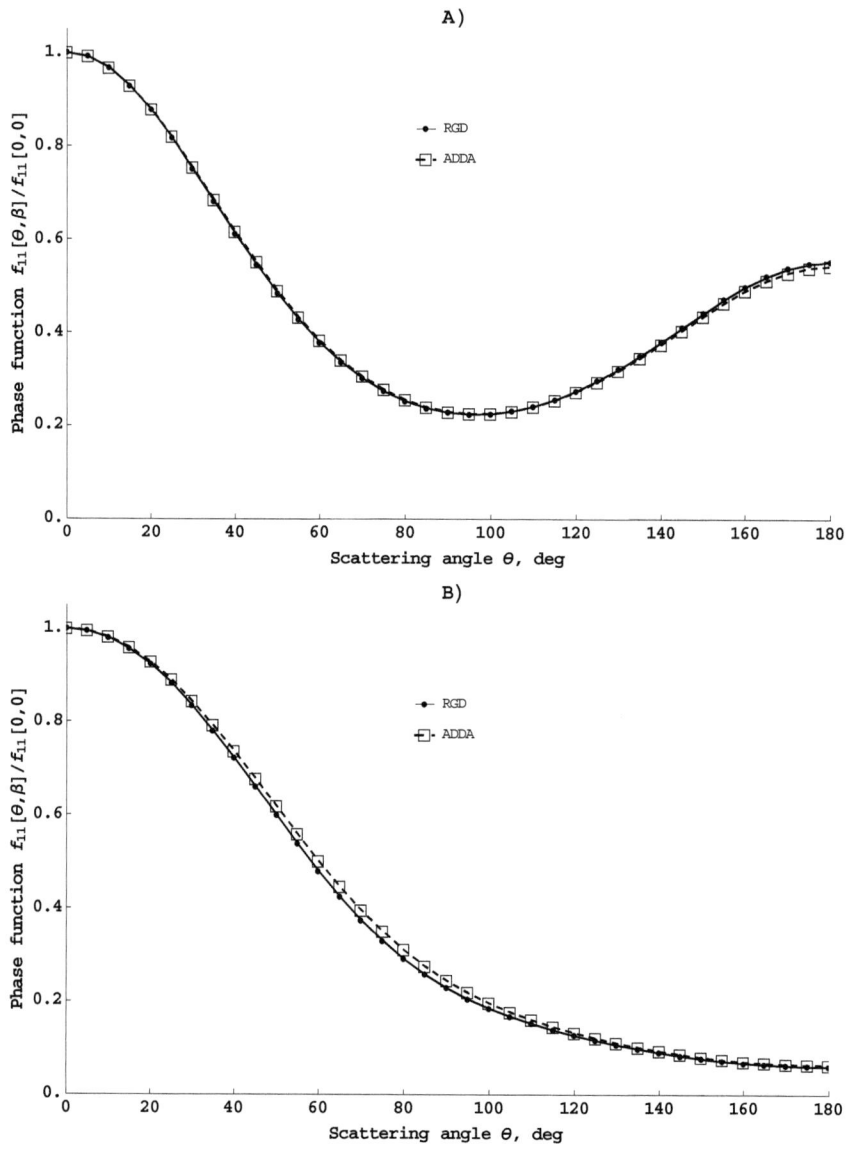

Fig. 15. Normalized phase function $f_{11}(\theta)/f_{11}(0)$ vs. scattering angle θ for a cone in the RGD approximation and ADDA with kR =2, kH =2 and direction of incident light along axis of symmetry β =0 (A) and perpendicular β =π/2 (B).

6. Polygonal pyramid

By analogy with prisms analytic equations for the light scattering amplitude in the RGD approximation of a pyramid consisting of an arbitrary polygonal base may be obtained using Eqs. (3),(5) [19, 20].

a)   b)

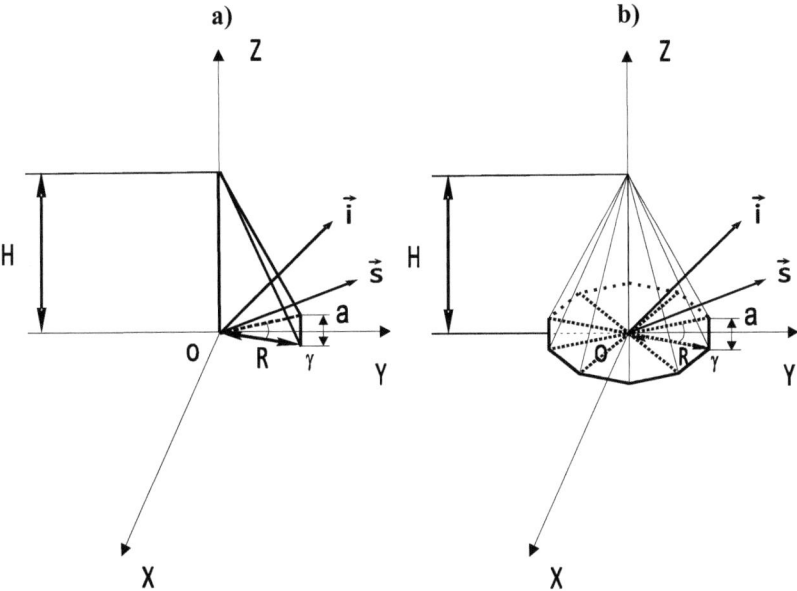

Fig. 16. Geometry of light scattering by a wedge of polygonal pyramid (a) and whole polygonal pyramid (b) [19].

The amplitude for polygonal pyramid segment wedge (see Fig. 16 a)) [19] is equal

$$f_{PD} = \frac{k^2(m^2-1)3V_{PD}}{4\pi k_1 k_3 RH \sin\gamma} \left[U(k_6 R, 0) - U(k_5 R, 0) + U(k_5 R, k_3 H) - U(k_6 R, k_3 H) \right], \qquad (18)$$

where $V_{PD} = \dfrac{1}{6} HR^2 \sin\gamma$ is a volume of pyramid wedge,

$U(x,y) = \dfrac{\exp(i x) - \exp(i y)}{x - y}$.

Rotating about axis OZ n-1 times light scattering amplitude for wedge pyramid (18) and summing all terms, we obtain for whole pyramid

$$f_{PYR} = \sum_{s=0}^{n-1} f_{PD}(s\,\gamma). \qquad (19)$$

Thus, light scattering amplitude for square base ($n=4$, $\gamma=\pi/2$) pyramid from Eq. (19) yield us

$$f_{PYR4} = \frac{k^2(m^2-1)V_{PD}}{2\pi} \frac{3\exp(i\,k_3H)}{k_1k_2R^2}[f_3+if_4],$$ (20)

where

$$f_3 = j_0(C_1)+j_0(C_2)-j_0(C_3)-j_0(C_4),\ f_4 = h_0(C_1)+h_0(C_2)-h_0(C_3)-h_0(C_4),$$

$$C_1 = k_3H+R^*(k_2-k_1),\qquad C_2 = k_3H-R^*(k_2-k_1),\qquad C_3 = k_3H+R^*(k_2+k_1),$$

$$C_4 = k_3H-R^*(k_2+k_1),\ R^* = R\big/\sqrt{2}.$$

And light scattering phase functions calculated using Eq. (21) in the RGD approximation and in the ADDA for particles with relative refractive index $m=1.1+i\cdot0.01$ are plotted in Fig. 17.

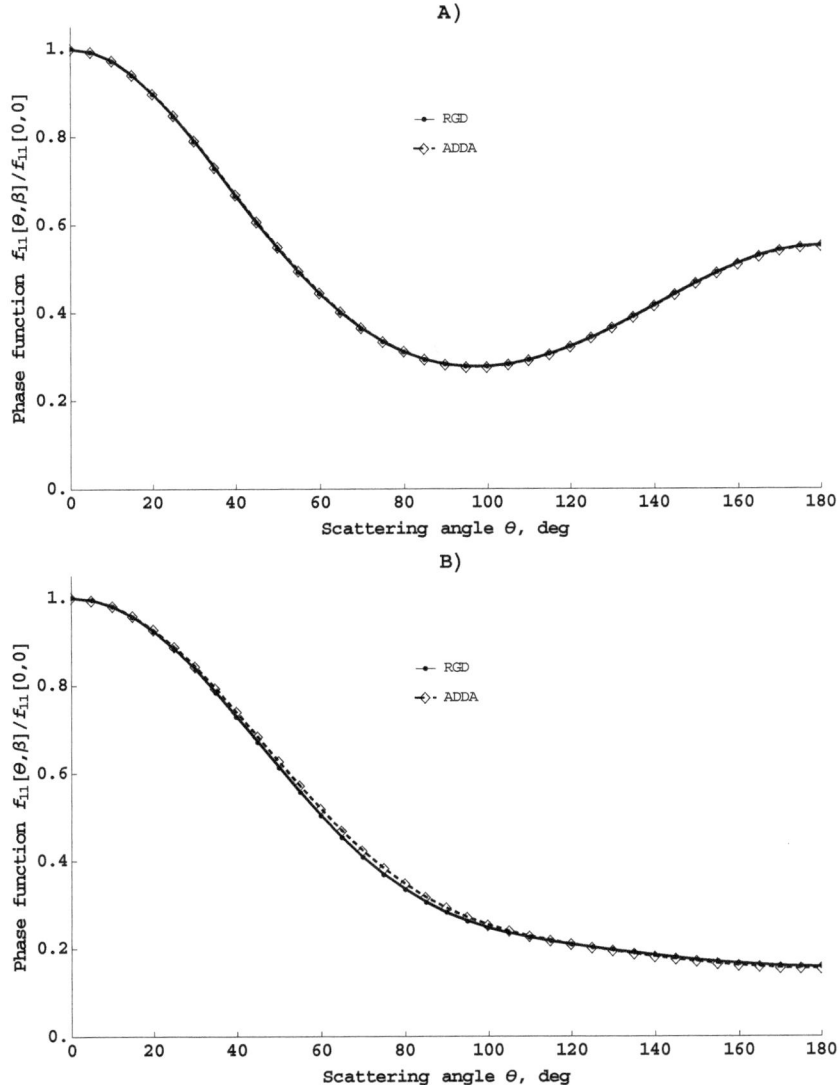

Fig. 17. Normalized phase function $f_{11}(\theta)/f_{11}(0)$ vs. scattering angle θ for a pyramid in the RGD approximation and ADDA with kR = 2, kH = 2, n = 4 and direction of incident light along axis of symmetry $\beta = 0$ (A) and perpendicular $\beta=\pi/2$ (B).

7. Torus

More complicate to find even light scattering amplitude for toroidal shape in the RGD approximation [10].

For an optically "soft" homogeneous tube (ring) of height and thickness 2a and radius R (Fig. 18a):

$$f_{TUBE} = \frac{k^2 a (m^2 - 1)}{k_4} \cdot [J_1(k_4(R+a))(R+a) - J_1(k_4(R-a))(R-a)] j_0(k_3 a), \qquad (21)$$

where $k_4 = \sqrt{k_1^2 + k_2^2}$.

Similarly, from Eq. (1), the light scattering amplitude will be obtained in a scalar form for a homogeneous torus with radii R and a (Fig. 18b):

$$f_{TORUS} = \frac{k^2 a (m^2 - 1)}{2k_4} \cdot \int_{-1}^{1} \cos(k_3 aq)[J_1(k_4(R+q_s))(R+q_s) - J_1(k_4(R-q_s))(R-q_s)] dq, \qquad (22)$$

where $q_s = a\sqrt{1 - q^2}$.

As a result, a somewhat cumbersome expression is obtained; therefore, we will give only an approximate consequence from Eq. (22) for the scattering angles α and β equal to or smaller than 90° (α = θ$_s$ – θ$_i$, β = φ$_s$ – φ$_i$) as follows:

$$f_{TI} = \frac{\pi k^2 a (m^2 - 1)}{2k_4} \cdot [J_1(k_4(R+a))(R+a) - J_1(k_4(R-a))(R-a)] \frac{J_1(k_3 a)}{k_3 a}. \qquad (23)$$

For small RGD toroidal particles provided kR≤1 and scattering angles α not more than 90⁰ modulus of relative error approximate light scattering phase functions is not more than 14%. Other features about Eq. (23) see in Ref. [10].

Some values of light scattering phase functions calculated using Eq. (22) for particle with relative refractive index m=1.1+i·0.01 are plotted in Fig. 19.

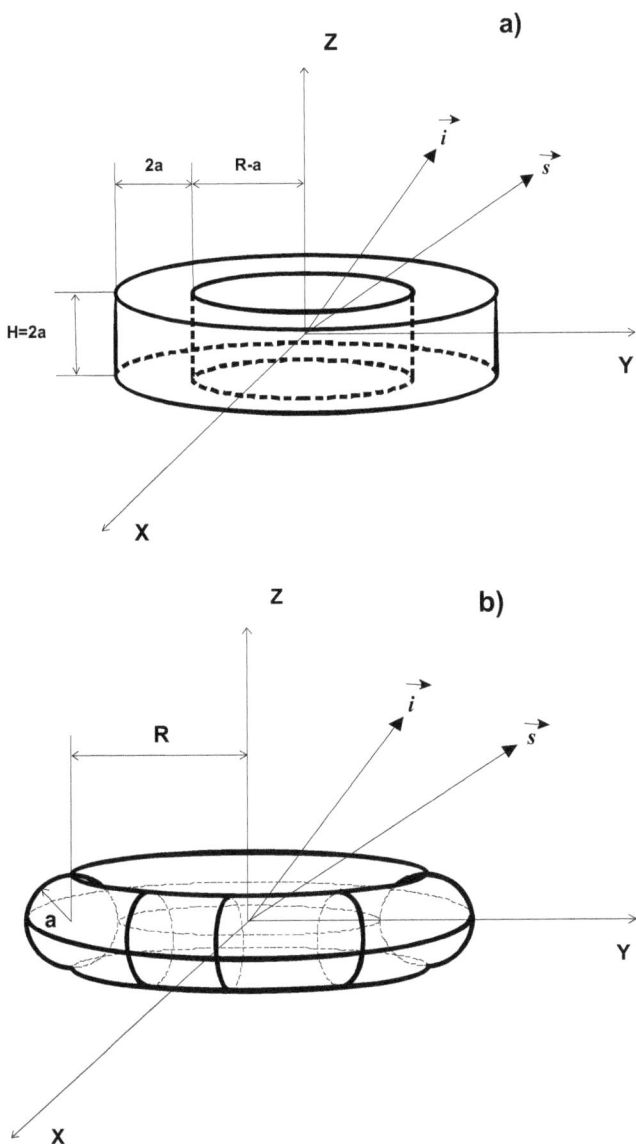

Fig. 18. Geometry of light scattering by a tube (a) and torus (b) [10].

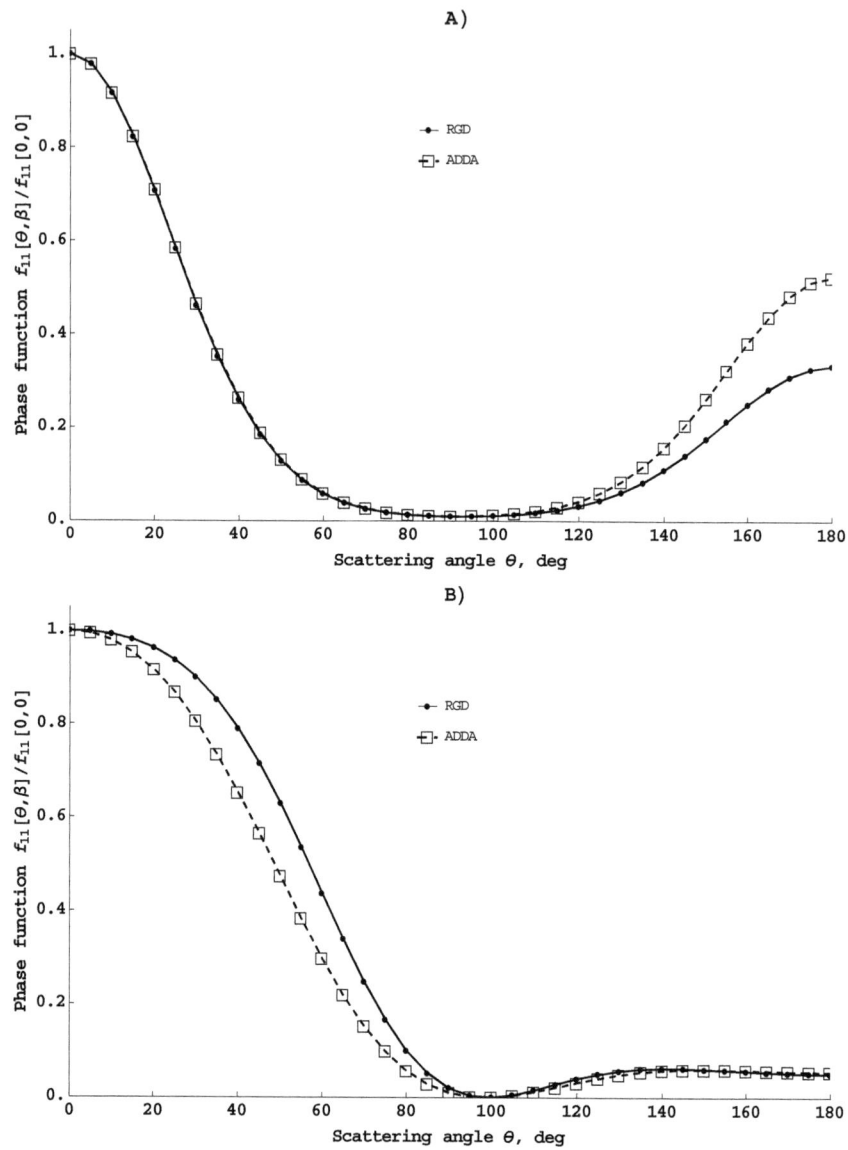

Fig. 19. Normalized phase function $f_{11}(\theta)/f_{11}(0)$ vs. scattering angle θ for a torus in the RGD approximation and ADDA with $kR = 2$, $kH = 2$, $ka=1$ and for direction of incident light along axis of symmetry $\beta = 0$ (A) and perpendicular $\beta = \pi/2$ (B).

8. Shell or hollow double layer particle

For hollow ($m_1=1$) double layered particle, having external form factor $\Phi_2(\theta,\beta)$, and internal form factor $\Phi_1(\theta,\beta)$, from Eq. (3) light scattering amplitude (see Ref. [27]) may be written as

$$f(\theta,\beta) = \frac{k^2(m^2-1)}{4\pi}[V_2\Phi_2(\theta,\beta) - V_1\Phi_1(\theta,\beta)]. \tag{24}$$

The light scattering amplitude for a homogeneous parallelepiped in the RGD approximation is a well-known [1, 4]:

$$f_P = \frac{k^2(m^2-1)V_P}{4\pi}j_0\left(\frac{k_1a}{2}\right)j_0\left(\frac{k_2b}{2}\right)j_0\left(\frac{k_3c}{2}\right), \tag{25}$$

where $V_P=abc$ is a volume of parallelepiped.

Usually, for thin layered objects the general expression (24) is used. So, for hollow thin layered parallelepiped (see Fig. 20 b)) from Eqs. (24), (25) the light scattering amplitude (as in [27]) follows

$$f_{TP} = \frac{k^2(m^2-1)}{4\pi}\left[V_{P2}j_0\left(\frac{k_1a_2}{2}\right)j_0\left(\frac{k_2b_2}{2}\right)j_0\left(\frac{k_3c_2}{2}\right) - V_{P1}j_0\left(\frac{k_1a_1}{2}\right)j_0\left(\frac{k_2b_1}{2}\right)j_0\left(\frac{k_3c_1}{2}\right)\right], \tag{26}$$

where $V_{P1}=a_1b_1c_1$ and $V_{P2}=a_2b_2c_2$ are volmes of internal and external parallelepiped respectively.

But, also the light scattering amplitude for thin layered parallelepiped may be obtained from Eqs. (3), (4) by summing amplitudes for 3 pairs of parallel planes (Fig. 20 a)) [27]:

$$f_{TP3} = \frac{k^2(m^2-1)}{4\pi}(f_5 + f_6 + f_7), \tag{27}$$

where $f_5 = (a_2 - a_1)b_1c_1\cos\left(\frac{k_1(a_1+a_2)}{4}\right)j_0\left(\frac{k_2b_1}{2}\right)j_0\left(\frac{k_3c_1}{2}\right)$,

$f_6 = (b_2 - b_1)a_2c_1\cos\left(\frac{k_2(b_1+b_2)}{4}\right)j_0\left(\frac{k_1a_2}{2}\right)j_0\left(\frac{k_3c_1}{2}\right)$,

$f_7 = (c_2 - c_1)a_2b_2\cos\left(\frac{k_3(c_1+c_2)}{4}\right)j_0\left(\frac{k_1a_2}{2}\right)j_0\left(\frac{k_2b_2}{2}\right)$.

(a)

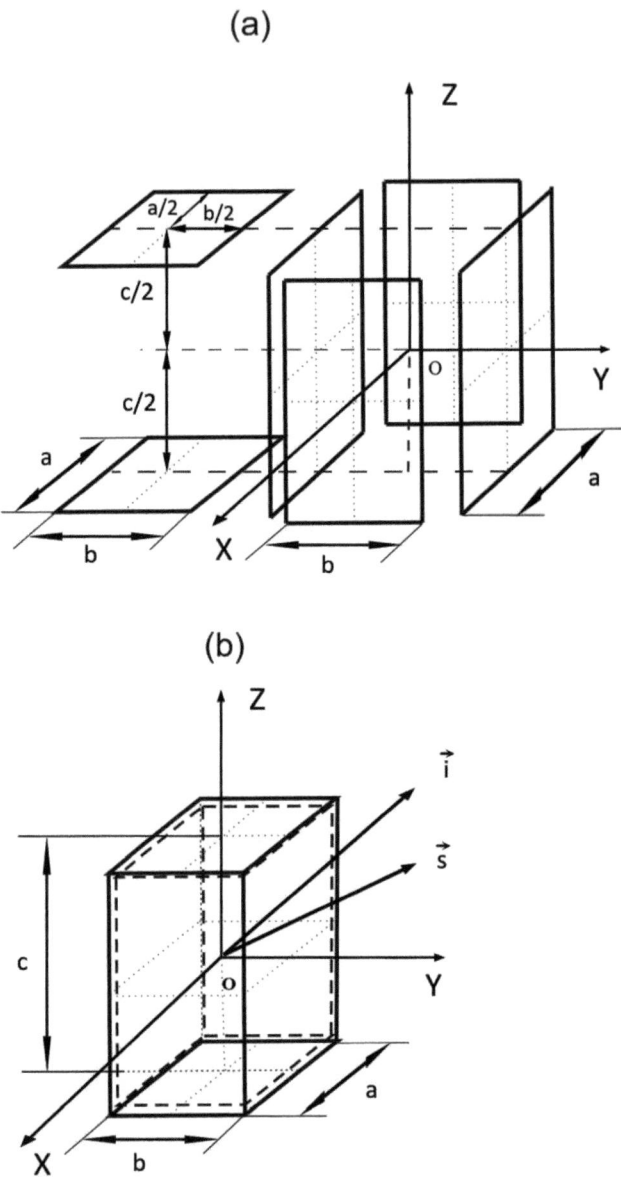

(b)

Fig. 20. Geometry of light scattering by a thin layered parallelepiped and formation from elements [27].

The advantage Eq. (27) over (26) is some stable under calculations and easy generalization for infinitely thin parallelepipeds.

For example, for the infinitesimally thin cube with edge a=b=c form factor follows from (27):

$$\Phi_{ITP} = \frac{\cos\left(\frac{k_1 a}{2}\right)j_0\left(\frac{k_2 a}{2}\right)j_0\left(\frac{k_3 a}{2}\right) + \cos\left(\frac{k_2 a}{2}\right)j_0\left(\frac{k_1 a}{2}\right)j_0\left(\frac{k_3 a}{2}\right) + \cos\left(\frac{k_3 a}{2}\right)j_0\left(\frac{k_1 a}{2}\right)j_0\left(\frac{k_2 a}{2}\right)}{3}. \tag{28}$$

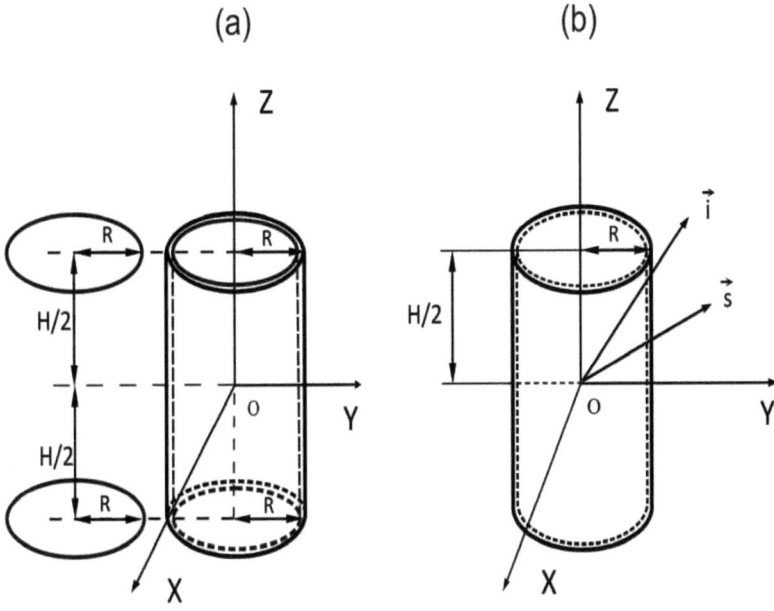

Fig. 21. Geometry of light scattering by a thin layered cylindrical capsule and formation from elements [27].

The light scattering amplitude for thin layered cylindrical capsule (see Fig. 21) may be easy obtained from general Eqs. (3), (4), (9) summing amplitudes for thin tube wall f_8 [4] and two thin parallel facing disks f_9 (see Ref. [27]):

$$f_{TC} = \frac{k^2(m^2-1)}{4\pi}(f_8 + f_9), \tag{29}$$

where $f_8 = \pi(R_2^2 - R_1^2)H_2 J_0\left(k_4\left(\frac{R_1+R_2}{2}\right)\right)j_0\left(\frac{k_3 H_2}{2}\right)$,

$$f_9 = \pi R_1^2 (H_2 - H_1) 2 \frac{J_1(k_4 R_1)}{k_4 R_1} \cos\left(k_3\left(\frac{H_1 + H_2}{4}\right)\right).$$

The expression (29) may be simplify for an infinitesimally thin layered cylindrical capsule, if R_1, $R_2 \to R$ and H_1, $H_2 \to H$, then we obtain form factor for infinitesimally thin layered cylindrical capsule:

$$\Phi_{ITC} = \frac{(f_8 + f_9)}{\Delta V}, \tag{30}$$

where $\dfrac{f_8}{\Delta V} = \dfrac{H}{R+H} J_0(k_4 R) j_0\left(\dfrac{k_3 H}{2}\right)$, $\dfrac{f_9}{\Delta V} = \dfrac{R}{R+H} \dfrac{2 J_1(k_4 R)}{k_4 R} \cos\left(\dfrac{k_3 H}{2}\right)$.

Note, the expression is similar (30), obtained earlier in [28] with error and corrected in [29].

Light scattering phase functions calculated using Eqs. (26)-(30) for particles with relative refractive index m=1.1+i·0.01 are shown in Figs. 22, 23.

Some researchers [30] have been investigated the range of validity of the RGD approximation for homogeneous nonspherical scatterers by performing scattering calculations for the case of two-layered spheroids using EBCM and the generalized point-matching technique (GPMT). It should be pointed out that although the maximum errors in the RGD approximation occur at scattering angles corresponding to the deep nulls in the differential scattering pattern (the RGD approximation invariably yields zero for these minima), however the positions of these nulls are very well predicted.

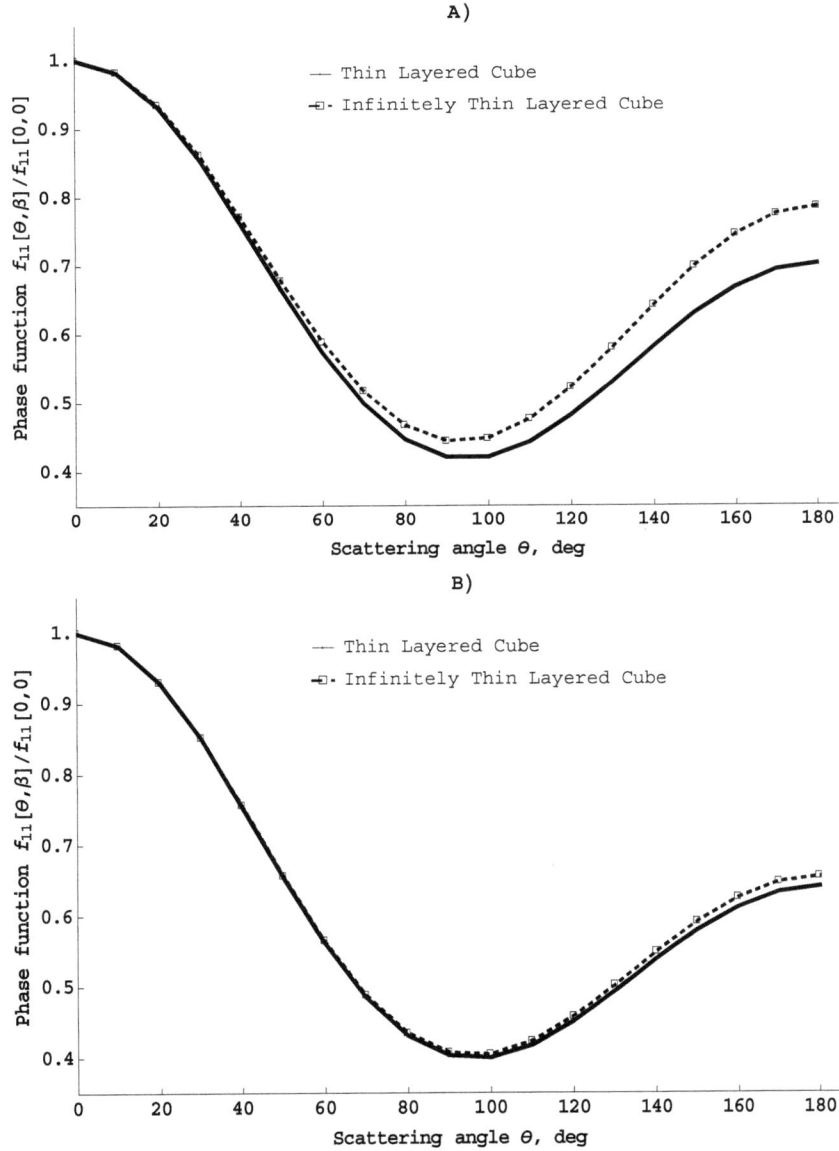

Fig. 22. Normalized phase function $f_{11}(\theta)/f_{11}(0)$ vs. scattering angle θ for thin layered cube and infinitesimally thin layered cube with $ka_2=kb_2=kc_2=1$ and $\theta_1=90^0$: A) $ka_1=kb_1=kc_1=0.3$; B) $ka_1=kb_1=kc_1=0.7$.

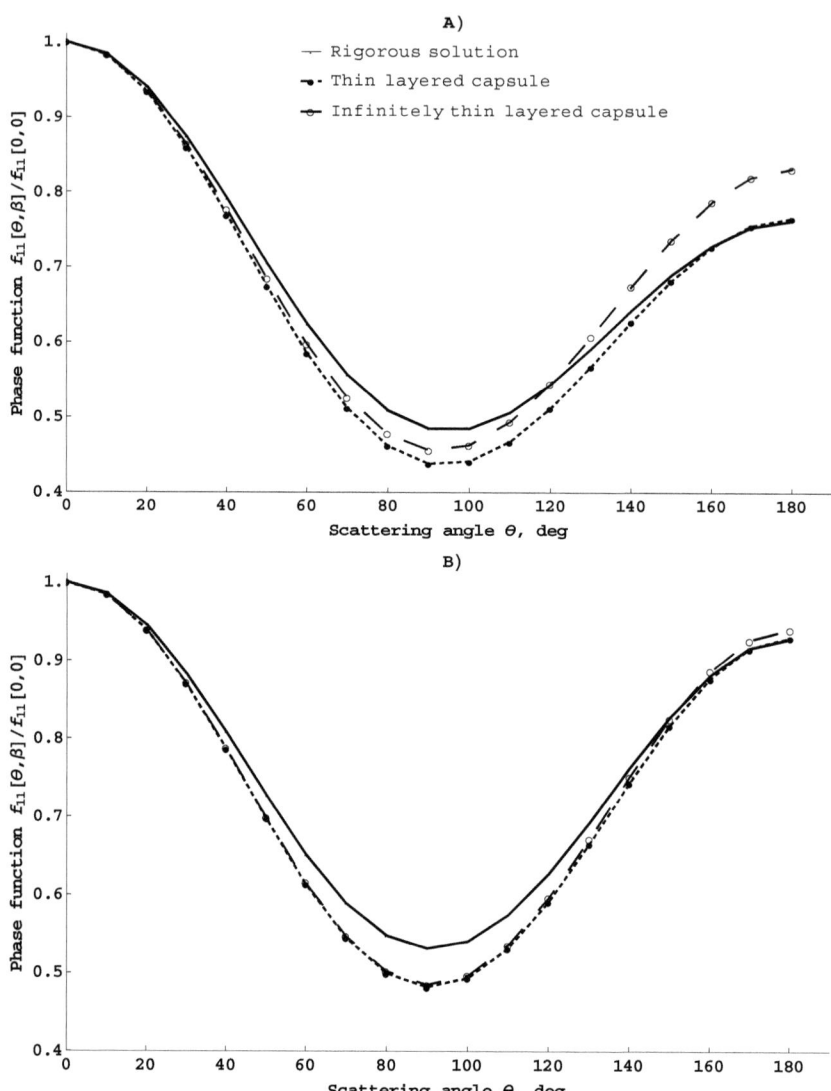

Fig. 23. Normalized phase function $f_{11}(\theta)/f_{11}(0)$ vs. scattering angle θ for thin layered cylindrical capsule and infinitesimally thin layered cylindrical capsule in comparison with infinitely long thin layered cylinder (rigorous solution) provided $kH_1=1.0$, $kH_2=1.01$, $kR_1=0.1$ and $\theta_i=90^0$: A) $kR_2=0.5$; B) $kR_2=0.25$.

9. Aggregates of particles

9.1 Strict oriented particles

For instance, let us assume that a system of three arbitrary identical particles are illuminated along OY axis (see, for example, circular cylinder paralleled OZ axis Fig. 24 [19]. Using Eqs. (4), (9) first cylinder of system located in the center of coordinates has the amplitude of light scattering f_{CYL}, two others disposed on the same distance d along OX axis to the left and to the right of first have the amplitude of light scattering $\exp(-ik_1d)f_{CYL}$ and $\exp(ik_1d)f_{CYL}$, respectively. Thus, using Eq. (3), the amplitude of a system of three identical cylinders is

$$f_{3CYL} = (1 + 2\cos(k_1 d))f_{CYL}. \tag{31}$$

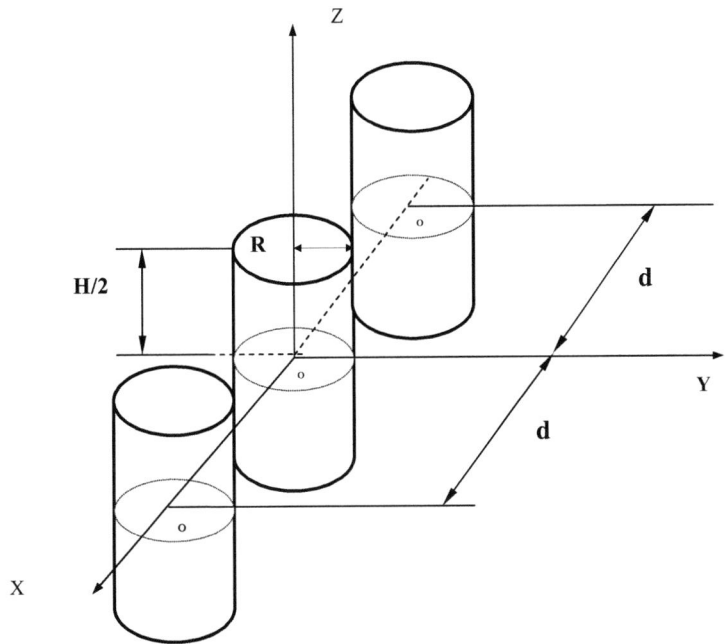

Fig 24. Geometry of light scattering by a system of three cylinders [19].

Using Eqs.(3),(4), we are generalized Eq.(31) for centered aggregate or ensemble consist of $N_X \times N_Y \times N_Z$ grid of identical particles each with form factor f_0:

$$f_{aggr} = f_0 \left(\sum_{p=0}^{Nx-1} \exp(ik_1 x_p) \right) \left(\sum_{q=0}^{Ny-1} \exp(ik_2 y_q) \right) \left(\sum_{s=0}^{Nz-1} \exp(ik_3 z_s) \right), \tag{32}$$

where x_p, y_q, z_s – cartesian coordinates of center of identical particles.

9.2 Randomly oriented particles

The average intensity of scattering by an ideal solution or disperse system is simply the sum of the intensities scattered by each individual molecule or particle, as, *ex hypothesi*, there is no interference between waves scattered by different particles. For the calculation of average intensity it is convenient to find average form factor for the randomly oriented particles in the RGD domain

$$\langle \Phi^2(\theta) \rangle = \frac{1}{4\pi} \int_\Omega |\Phi|^2 d\Omega \,, \tag{33}$$

where the values of θ (and φ) are considered fixed and the integration over the solid angles Ω refers to the orientations that the particles can have.

If particle have an axis of symmetry then we can rewrite (33) as

$$\langle \Phi^2(\theta) \rangle = \frac{1}{4\pi} \int_0^{2\pi} \int_0^\pi |\Phi|^2 \sin\beta \, d\beta \, d\phi \,. \tag{34}$$

For a single thin rod of length L [2, 4] from Eq.(34) average form factor has been found

$$\langle \Phi^2(\theta) \rangle = \frac{1}{Z} \int_0^{2Z} j_0(w) \, dw - [j_0(Z)]^2 \,, \tag{35}$$

where $Z = k_S L$.

For a single thin disk of radius R [2, 4] from Eq.(34) average form factor has been obtained

$$\langle \Phi^2(\theta) \rangle = \frac{2}{Z^2} \left\{ 1 - \frac{J_1(2Z)}{Z} \right\} \,, \tag{36}$$

where $Z = 2 k_S R$.

The simplest type of aggregate to consider is that in which the density of scattering material is distributed in a spherically symmetrical fashion.

For such N randomly oriented spheres the averaged intensity of scattering has been obtained in [31, 32]

$$\left(\frac{3 j_1(k_S R)}{k_S R} \right)^2 \frac{1}{N^2} \sum_{i=1}^N \sum_{j=1}^N \frac{\sin(k_S r_{ij})}{k_S r_{ij}} \,, \tag{37}$$

where r_{ij} is the distance between the ith and jth electron.

Also for aggregates of N parallel infinitely long rods with fixed locations the averaged intensity of scattering has been obtained earlier [33]:

$$\left(\frac{2J_1(k_sR)}{k_sR}\right)^2 \frac{1}{N^2}\sum_{i=1}^{N}\sum_{j=1}^{N} J_0\left(k_s r_{ij}\right), \tag{38}$$

where r_{ij} is the distance between the centers of ith and jth cylinders.

For seven infinitely long cylinders in a central hexagonal arrangement [33] (see geometry in Fig. 25), the expression for the scattered intensity becomes

$$\left(\frac{2J_1(k_sR)}{k_sR}\right)^2 \frac{1}{49}\left(7 + 24J_0(x) + 6J_0(2x) + 12J_0\left(\sqrt{3}x\right)\right), \tag{39}$$

where $x = k_s r$, r is a radius or side of hexagon.

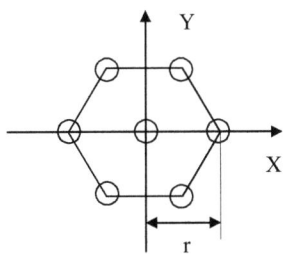

Fig. 25. Geometry of seven cylinders in a central hexagonal arrangement.

See also Appendix Table 2.

If single particle is randomly oriented, then light scattering efficiency factor Q_S^R (without consideration of symmetry on azimuthal angle ϕ) can be defined as

$$Q_S^R = \frac{\int\limits_0^{2\pi}\int\limits_0^{\pi} Q_S(\theta_i,\phi_i)\cdot S(\theta_i,\phi_i)\cdot\sin(\theta_i)d\theta_i d\phi_i}{\int\limits_0^{2\pi}\int\limits_0^{\pi} S(\theta_i,\phi_i)\cdot\sin(\theta_i)d\theta_i d\phi_i}. \tag{40}$$

Substituting areas Eqs. (11) and (15) into Eq. (40), we have found denominator of light scattering efficiency factor Q_S^R for circular cylinder it gives $2\pi^2 \cdot RH$, for hexagonal cylinder respectively $6\pi \cdot RH$ [25].

If size of particle is small $kR<1$ and $kH<1$ then Eq. (40) may be expanded in Taylor series [25], and for the random oriented small circular cylinder in the RGD approximation we can write in scalar form

$$Q_S^R = \frac{(kR)^3(kH)\left(n^2-1\right)^2}{24}\left\{8 - \frac{22}{45}(kH)^2 - \frac{38}{15}(kR)^2 + \frac{13}{105}(kR)^2(kH)^2\right\}. \tag{41}$$

Also for the random oriented small hexagonal cylinder in the RGD approximation [25] light scattering efficiency factor Q_S^R has been obtained in scalar form

$$Q_S^R = \frac{3(kR)^3(kH)(n^2-1)^2}{32\pi} \cdot \left\{ 8 - \frac{22}{45}(kH)^2 - \frac{76}{45}(kR)^2 + \frac{26}{315}(kR)^2(kH)^2 \right\}. \quad (42)$$

Further, the comparison of numerical calculations of light scattering efficiency factor Q_S^R obtained with formula (40) in the RGD approximation and the method of ADDA (or DDA [23]) have been performed. The results of numerical calculations of light scattering efficiency factor for randomly oriented circular and hexagonal cylinders using Eqs.(41),(42) (corrected with polarization matrix proposed in [34]) and by discrete dipole method with relative index of refraction m=1.31+i·0.01 are shown in Figs. 26, 27.

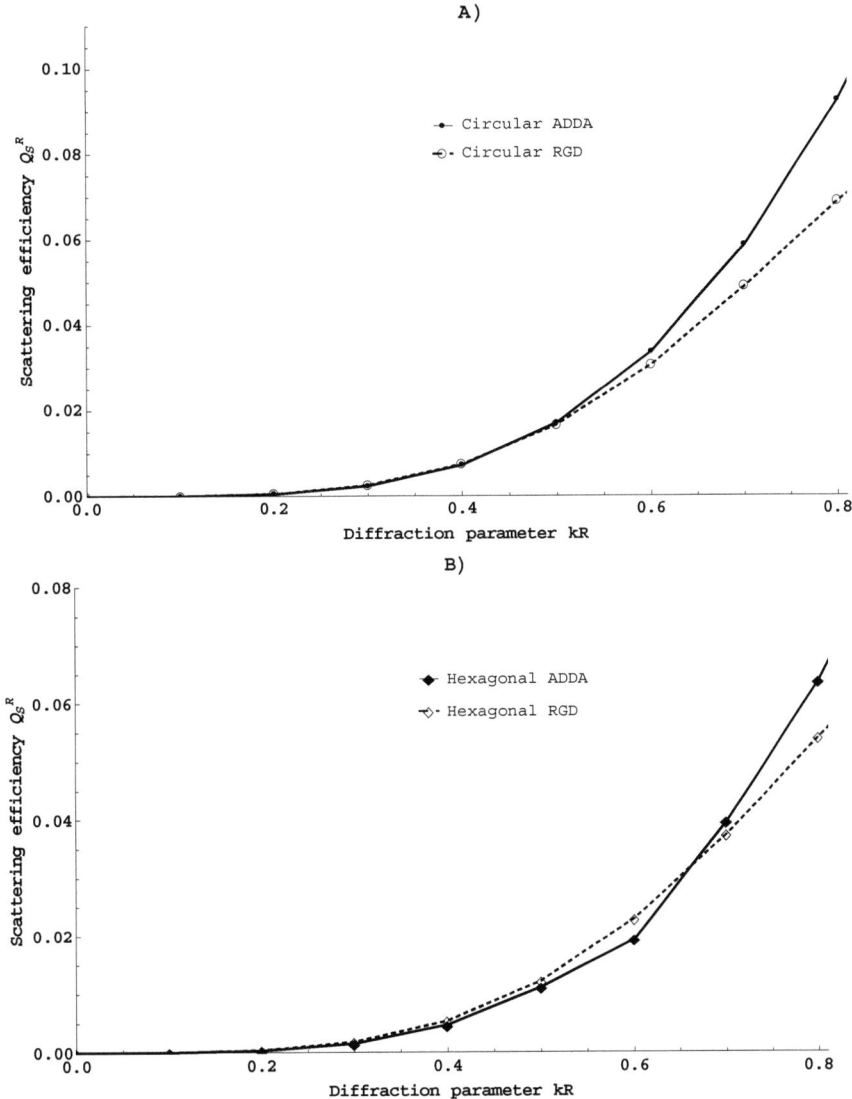

Fig. 26. Light scattering efficiency Q_S^R vs. diffraction parameter kR in the RGD approximation and ADDA for randomly oriented circular cylinders and hexagonal cylinders with relative refractive index m=1.31+i·0.01 provided aspect ratio H/(2R) = 3/2.

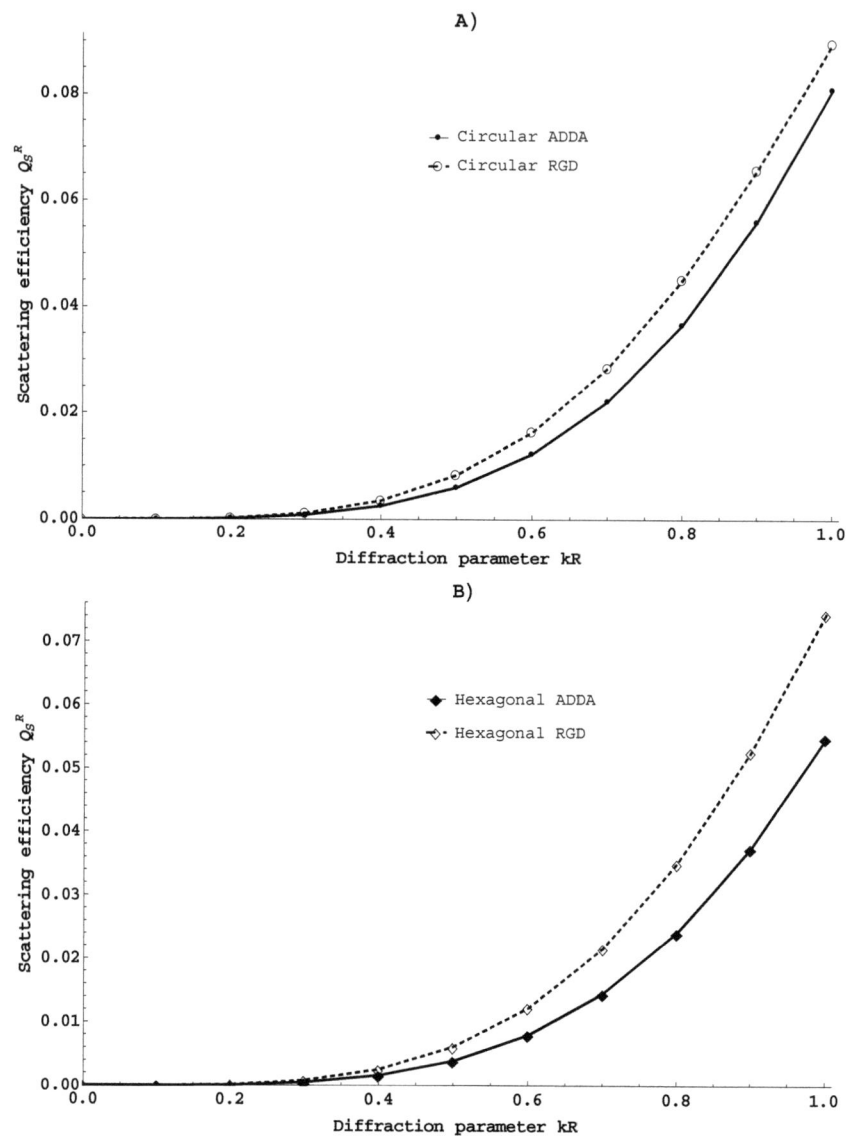

Fig. 27. Light scattering efficiency Q_S^R vs. diffraction parameter kR in the RGD approximation and ADDA for randomly oriented circular cylinders and hexagonal cylinders with relative refractive index m=1.31+i·0.01 provided aspect ratio H/(2R) = 2/3.

For instance, absolute value of relative error of the calculation of light scattering efficiency factors Q_S^R, calculated with such approximate expressions without polarization correction, in comparison with calculation by the DDA under kR<1 and kH<1 and ratio height to radius from 0.5 to 2 for circular cylinder is not more than 17%, and for hexagonal cylinders respectively 13% [25].

9.3. Polydisperse particles

In an ensemble of particles the effect of nonsphericity is not significant for small particles. If the particle size is such that the Rayleigh approximation is applicable, it is found that the light scattered by randomly oriented spheroids is almost indistinguishable from light scattered by a collection of spheres of equal volume [35].

On the question of equivalence, further work has been done within the framework of the Rayleigh approximation in [36]. They have derived a theorem for the optical properties of particles of arbitrary shapes. The further theorem is stated in [11, 36].

Theorem. *The absorption cross-section of an arbitrarily shaped and arbitrarily oriented, homogeneous particle in the Rayleigh domain, or an ensemble of such particles with various shapes and orientations, equals the average cross-section of an ensemble of spheroidal particles in a fixed orientation with the same composition and a shape distribution that is independent of the composition of the particle.*

Generally, the two approaches are well known to solve a problem of interaction light with random oriented nonspherical particles of intermediate sizes [8].

First approach is the description of optical characteristics every particle and then obtaining of average result. This approach is very hard and complicated even under modern developing of computer technologies.

Second approach is the substitution of a system of nonspherical particles for some average particle [37-41]. The model of ensemble of nonspherical particles is widely practised "equivalent" spheres, i.e. spheres with equal volume (V) or equal area of surface (S) which having nonspherical particles. (index of refraction is a same too). However, such methods are allowed to estimate characteristics of light scattering only in first approximation. Really, equivolume spheres have too small area of surface, that leads to unsatisfied description of characteristics of light scattering, but equisurface spheres have too large volume, that, in this case, leads to overestimate of the absorption. One can significantly improve description of characteristics of light scattering by the substitution of ensemble of nonspherical particles for set (ensemble) of spherical particles, which at same time have equal volume and area of surface, that have nonspherical too.

Earlier it has been shown that representation of random oriented nonspherical particles as ensemble of spherical particles V/S sphere follow us to more precise determination of integral characteristics of light scattering of ensemble of nonspherical particles [42-45]. So, for example, in [37-39] for description of integral characteristics of light scattering of hexagonal prisms the ensemble of identical spheres is effectively used. Under this approach for alone nonspherical particle formally is corresponded several spherical particles, which number determined from condition of "equivalency" of volume and area of surface.

Obviously, that the condition of "equivalency" of volume and area of surface may be obtained as substitution of nonspherical particle for ensemble of spherical particles with some size distribution, moreover, there are infinite numbers of such distributions.

Full analytically proof of the application of model V/S sphere for the description of field, scattered by suspension of optically "soft" random oriented nonspherical particles is presented in Refs. [8, 44, 45]. There are brief steps of this proof with some redefined and reassigned variables. In the RGD approximation for homogeneous ellipsoid of revolution the intensity $I(\theta)$ of light scattered at angle to the incident beam is given by expression [4, 8, 44, 45] (here authors use condition of optically "soft" particle as $m^2-1 \approx 2(m-1)$):

$$I(\theta) = \frac{k^4 V^2 |m-1|^2}{8\pi^2 r^2} I_0 G^2(U)(1 + \cos^2 \theta),$$ (43)

where V is the volume of particle, $G(U) = \dfrac{3 j_1(U)}{U} = \dfrac{3}{U^3}(\sin U - U \cos U)$,

$Z = \cos\beta$, $U = 2kc \sin\left(\dfrac{\theta}{2}\right)\sqrt{\varepsilon^2 - (\varepsilon^2 - 1)Z^2}$, $j_1(x)$ is a spherical Bessel function of first order, I_0 is the intensity of the incident light, r is the distance to the detector, $\varepsilon = R/c$ is the axial ratio, c is semisize of axis of revolution.

In the case of $\varepsilon = 1$ Eq. (43) reduces to the well-known expression for the light-scattering intensity of a homogeneous sphere where $U = 2kc \sin(\theta/2)$.

For randomly oriented particles the intensity $I(kc, \theta)$ is

$$I(kc, \theta) = \frac{(1 + \cos^2 \theta)|m-1|^2}{2k^2 r^2} \frac{4}{9} I_0 (kc)^6 \varepsilon^4 \int_0^1 G^2(U) dZ.$$ (44)

Eq. (44) can be interpreted as scattering by polydisperse ensemble of spheres:

$$I(kc,\theta) = \frac{\left(1+\cos^2\theta\right)|m-1|^2}{2k^2r^2} \frac{4}{9} I_0 \int_0^1 \frac{(kc)^6 \varepsilon^4 \alpha_S^6}{\alpha_S^6} G^2(U)dZ \; , \qquad (45)$$

where $\alpha_S = kc\sqrt{\varepsilon^2 - (\varepsilon^2 - 1)Z^2}$, $U = 2\alpha_S\sin(\theta/2)$.

Using a substitution of variables, α_S, we can write Eq. (45) as in [44, 45]

$$I(kc,\theta) = \int_{kc}^{kR} I(\alpha_S, \theta) \frac{(kR)^4 kc}{(\varepsilon^2 - 1)\alpha_S^5} \sqrt{\frac{\varepsilon^2 - 1}{(kR)^2 - \alpha_S^2}} d\alpha_S \; . \qquad (46)$$

Eq. (46) shows that the optical properties of the randomly oriented spheroids are identical to optical properties of a system of polydisperse spheres with the power law size distribution [44, 45], which is as follows:

$$n(\alpha_S) = \frac{(kR)^4 kc}{(\varepsilon^2 - 1)\alpha_S^5} \sqrt{\frac{\varepsilon^2 - 1}{(kR)^2 - \alpha_S^2}} = \frac{(kR)^4 kc}{\alpha_S^5 \sqrt{(\varepsilon^2 - 1)((kR)^2 - \alpha_S^2)}} \; . \qquad (47)$$

The maximal and minimal size parameters α_S in this distribution correspond to maximal and minimal size parameters of spheroid.

This distribution has two important features. Namely, the averaged surface area and the volume of spheres in such distribution coincide with the surface area (S) and the volume (V) of a spheroid, i.e.

$$S = \int_{kc}^{kR} 4\pi a_S^2 n(\alpha_S)d\alpha_S \; , \quad V = \int_{kc}^{kR} \frac{4\pi}{3} a_S^3 n(\alpha_S)d\alpha_S \; , \qquad (48)$$

where $\alpha_S = k\, a_S$, a_S is the radius of a sphere.

Note also that Eq. (47) can be applied to describe scattering cross section and integral phase function of randomly oriented RGD spheroids [45].

10. Conclusions

Thus, we were discussed the general approach to obtaining of the form factor for a compound particle in the RGD approximation. The addition and rotational-translational properties of light scattering amplitude in the RGD approximation formulated herein allow us to construct the form factor for a system of several particles too. As a result of application of this technique formulas for the amplitude of light scattering by a prism and pyramid with an arbitrary polygonal base in the RGD approximation were obtained. The formulas obtained earlier for the amplitude of light scattering by a hexagonal cylinder and cone in the RGD approximation were presented too. Moreover our numerical calculations of phase functions and light scattering efficiency factors are successfully compared with ADDA methods.

In general, these expressions and technique may be also useful for the analytical evaluation of generalized parameters of polydisperse systems of particles, for the construction of new more precise approximations, for the comparisons with other solutions, techniques and for the further analysis of the validity's range of the RGD approximation.

Appendix

Table 1. Form factors for some particles with axis of symmetry [25]

Particle shape	Volume	$\Phi(\theta, \beta)$ for $k_4=0$	$\Phi(\theta, \beta)$
Sphere	$\dfrac{4}{3}\pi R^3$	$\dfrac{3\,j_1(k_3 R)}{k_3 R}$	$\dfrac{3\,j_1(k_S R)}{k_S R}$
Half Sphere	$\dfrac{2}{3}\pi R^3$	$\dfrac{3\left(j_1(k_3 R)+i\,h_1(k_3 R)\right)}{k_3 R}$	-
Spheroid	$\dfrac{4}{3}\pi R^2 c$	$\dfrac{3\,j_1(k_3 c)}{k_3 c}$	$\dfrac{3\,j_1\left(\sqrt{(k_4 R)^2+(k_3 c)^2}\right)}{\sqrt{(k_4 R)^2+(k_3 c)^2}}$
Cylinder	$\pi R^2 H$	$j_0\!\left(k_3\,\dfrac{H}{2}\right)$	$\dfrac{2J_1(k_4 R)}{k_4 R}\,j_0\!\left(k_3\,\dfrac{H}{2}\right)$
Infinite cylinder $\theta i=\pi/2$ [7]	-	-	$\dfrac{2J_1(k_4 R)}{k_4 R}$
Cone	$\dfrac{\pi}{3}R^2 H$	$\dfrac{6}{k_3 H}\{h_0(k_3 H)-j_1(k_3 H)+$ $i(1-h_1(k_3 H)-j_0(k_3 H))\}$	provided $k_3 H=k_4 R=w$ $\dfrac{2\exp(iw)}{w}[q+ip]$, where $q=\cos w\,J_1(w)+\sin w\,J_2(w)$ $p=\cos w\,J_2(w)-\sin w\,J_1(w)$
Paraboloid of revolution	$\pi R^2 H\,\dfrac{p}{2}$	$\dfrac{2}{k_3 H}[h_0(k_3 H)+i(1-j_0(k_3 H))]$	-
Torus	$2\pi^2 R\,a^2$	$\dfrac{2J_1(k_3 a)}{k_3 a}$	provided $k_4 R<1$ $\dfrac{8J_1(k_4(R+a))}{k_4(R+a)}\dfrac{J_1(k_4(R-a))}{k_4(R-a)}\dfrac{J_1(k_3 a)}{k_3 a}$
Cassini oval based body $a>c$	$\dfrac{c^4\pi^2}{2a}$	$\dfrac{2J_1\left(\dfrac{k_3 c^2}{2a}\right)}{\dfrac{k_3 c^2}{2a}}$	-
Cassini oval based body $0<a\leq c$	$\dfrac{\pi}{3}\sqrt{c^2-a^2}\,(2c^2+a^2)+$ $\dfrac{\pi c^4}{a}\arcsin\left(\dfrac{a}{c}\right)$	-	-

Table 2. Form factors for randomly oriented particles

Particle of arbitrary shape	$$\langle \Phi^2(\theta) \rangle = \frac{1}{4\pi} \int_\Omega |\Phi|^2 d\Omega \,,$$ Φ is a form factor of strict oriented particle
Particle with axis of symmetry	$$\frac{1}{4\pi} \int_0^{2\pi}\int_0^\pi |\Phi|^2 \sin\beta \, d\beta \, d\phi$$
Spherical electron distribution in the polyatomic gases [31]	$$\left(\frac{3 j_1(k_S R)}{k_S R} \right)^2 \frac{1}{N^2} \sum_{i=1}^{N}\sum_{j=1}^{N} \frac{\sin(k_S r_{ij})}{k_S r_{ij}} \,, \text{ where } r_{ij} \text{ is the}$$ distance between the ith and jth electron.
Two spheres in contact [32]	$$\left(\frac{3 j_1(k_S R)}{k_S R} \right)^2 \frac{1}{4}\left(2 + \frac{\sin(2k_S R)}{2k_S R} \right)$$
Cylinder [2,4]	$$\int_0^{\pi/2} \left[\frac{2 J_1(k_S R \sin\beta)}{k_S R \sin\beta} j_0\left(k_S \frac{H}{2}\cos\beta \right) \right]^2 \sin\beta \, d\beta$$
Thin rod of length L [2,4]	$$\frac{1}{Z} \int_0^{2Z} j_0(w)\,dw - [j_0(Z)]^2 \,, \text{ where } Z = k_S L$$
Thin disk of radius R [2,4]	$$\frac{2}{Z^2}\left\{ 1 - \frac{J_1(2Z)}{Z} \right\} \,, \text{ where } Z = 2k_S R$$
Assembly of infinitely long cylinders [33]	$$\left(\frac{2 J_1(k_S R)}{k_S R} \right)^2 \frac{1}{N^2} \sum_{i=1}^{N}\sum_{j=1}^{N} J_0(k_S r_{ij}), \text{ where } r_{ij} \text{ is the}$$ distance between the centers of ith and jth cylinders.
Two infinitely long cylinders in contact [33]	$$\left(\frac{2 J_1(k_S R)}{k_S R} \right)^2 \frac{1}{4}(2 + 2 J_0(2k_S R))$$
Seven infinitely long cylinders [33]	$$\left(\frac{2 J_1(k_S R)}{k_S R} \right)^2 \frac{1}{49}(7 + 24 J_0(x) + 6 J_0(2x) + 12 J_0(\sqrt{3}x)),$$ where x= k_Sr, r is a radius or side of hexagon.

References

[1]   C.F. Bohren and D.R. Huffman, *Absorption and Scattering of Light by Small Particles*, John Wiley & Sons, New York, 1983.

[2]   H.C. van de Hulst, *Light Scattering by Small Particles,* John Wiley & Sons, New York, 1957.

[3]   M.I. Mishchenko, J.W. Hovenier, and L.D. Travis, *Light Scattering by Nonspherical Particles: Theory, Measurements, and Applications*, Academic Press, San Diego, 2000.

[4]   M. Kerker, *The Scattering of Light and Other Electromagnetic Radiation.* Academic Press, New York, London, 1969.

[5]   A.A. Kokhanovsky, *Cloud Optics,* Springer, Dordrecht, 2006.

[6]   V.V. Tuchin, ed., *Handbook of Optical Biomedical Diagnostics.* vol. PM107, SPIE Press, Bellingham, 2002.

[7]   Lord Rayleigh (J.W. Strutt), "On the Electromagnetic Theory of Light," *Phil. Mag.* Series 5, vol. 12, No. 73, pp. 81-101, 1881.

[8]   V.N. Lopatin, A.V. Priezzev, A.D. Aponasenko, N.V. Shepelevich, V.V. Lopatin, P.V. Pozhilenkova, and I.V. Prostakova, *Methods of Light Scattering in Analysis of Dispersion Biological Media,* PhysMatLit, Moscow, 2004. (in Russian).

[9]   P. W. Barber and D.-S. Wang, "Rayleigh-Gans-Debye applicability to scattering by nonspherical particles," *Appl. Opt.*, vol. 17, pp. 797-803, 1978.

[10]   K.A. Shapovalov, "Light scattering by particles of toroidal shape in the Rayleigh-Gans-Debye approximation," *Opt. Spectrosc.*, vol. 110, No.5, pp. 806-810, 2011.

[11]   S.K. Sharma and D.J. Somerford, *Light Scattering by Optically Soft Particles: Theory and Applications,* Springer-Praxis, Berlin, Heidelberg, New York, 2006.

[12]   P. Chýlek and J.P. Klett, "Extinction cross section of non-spherical particles in the anomalous diffraction approximation," *J. Opt. Soc. Am. A*, vol. 8, pp. 274-281, 1991.

[13]   P. Chýlek and J.P. Klett, "Absorption and scattering of electromagnetic radiation by prismatic columns: Anomalous diffraction approximation," *J. Opt. Soc. Am. A*, vol. 8, pp. 1713-1720, 1991.

[14]   V.M. Rysakov, "Light scattering by "soft" particles of arbitrary shape and size," *J. Quant. Spectrosc. Radiat. Transfer*, vol. 87, pp. 261-287, 2004.

[15]   K. Muinonen, "Light scattering by Gaussian random particles: Rayleigh and Rayleigh-Gans approximations," *J. Quant. Spectrosc. Radiat. Transfer*, vol. 55, pp. 603-613, 1996.

[16]   J.W. Shepherd and A.R. Holt, "The scattering of electromagnetic radiation from finite dielectric circular cylinders," *J. Phys. A: Math. Gen.,* vol. 16, pp. 651-662, 1983.

[17]   A.R. Holt, N.K. Uzunoglu, and B.G. Evans, "An integral equation solution to the scattering of electromagnetic radiation by dielectric spheroids and ellipsoids," *IEEE Trans. Antennas Propag.,* vol. 26, pp. 706-712, 1978.

[18]   A. Ishimaru, *Wave Propagation and Scattering in Random Media,* IEEE Press, New York, 1997.

[19]   K.A. Shapovalov, "Light Scattering by a Prism and Pyramid in the Rayleigh-Gans-Debye Approximation," *Optics,* vol. 2, No. 2, pp.32-37, 2013. - doi:10.11648/j.optics.20130202.11.
http://article.sciencepublishinggroup.com/pdf/10.11648.j.optics.20130202.11.pdf

[20]   K.A. Shapovalov, "Amplitude of Light Scattering by a Truncated Pyramid and Cone in the Rayleigh-Gans-Debye Approximation," *European Researcher,* vol.49, No.5-2, pp.1291-1297, 2013 (in Russian).

[21]   M.I. Mishchenko, "Calculation of the amplitude matrix for a nonspherical particle in a fixed orientation," *Appl. Opt.,* vol. 39, No. 6, pp. 1026-1031, 2000.

[22]   M.A. Yurkin and A.G. Hoekstra, "The discrete-dipole-approximation code ADDA: Capabilities and known limitations," *J. Quant. Spectrosc. Rad. Transf.,* vol. 112, pp.2234-2247, 2011.

[23]   B.T. Draine and P.J. Flatau, "Discrete-dipole approximation for scattering calculations," *J. Opt. Soc. Am. A.,* vol. 11, pp.1491-1499, 1994.

[24]   K.A. Shapovalov, "Light scattering of cylindrical particles in Rayleigh-Gans-Debye approximation. 1. Rigorously oriented particles," *Atmos. Oceanic. Opt.,* vol. 17, No. 4, pp. 350-353, 2004.

[25]   K.A. Shapovalov, "Light scattering of cylindrical particles in Rayleigh-Gans-Debye approximation. 2. Randomly oriented particles," *Atmos. Oceanic. Opt.,* vol. 17, No. 8, pp. 627-629, 2004.

[26]   K.A. Shapovalov, "Light scattering by particles with axis of symmetry in Rayleigh-Gans-Debye approximation," *J. Sib. Fed. Univ. Math. Phys.,* vol. 5, No. 4, pp. 586-592, 2012.   http://elib.sfu-kras.ru/bitstream/2311/3112/1/shapevalev.pdf (Open Access Journal in Russian).

[27]   K.A. Shapovalov, "Light scattering by thin shell particles in the Rayleigh-Gans-Debye approximation," *Innovacii i Investicii,* No. 3, pp. 244-247, 2014. (in Russian).

[28] J.S. Pedersen, "Analysis of small-angle scattering data from colloids and polymer solutions: modeling and least-squares fitting," *Adv. Colloid Interface Sci.*, vol. 70, pp.171-210, 1997.

[29] J.S. Pedersen et al., "Contrast variation small-angle neutron scattering study of the structure of block copolymer micelles in a slightly selective solvent at semidilute concentrations," *Macromolecules,* vol. 33, pp.542-550, 2000.

[30] J. M. Tranquilla and H. M. Al-Rizzo, "Electromagnetic Scatering from Dielectric-Coated Axisymmetric Objects Using the Generalized Point-Matching Technique (GPMT)," *IEEE Trans. Antennas Propagat.*, vol. 43, No. 1, 1995.

[31] P. Debye, "Zerstreuung von röntgenstrahlen,"*Ann. Phys.*, vol. 46, pp. 809-823, 1915.

[32] G. Oster and D.P. Riley, "Scattering from Isotropic Colloidal and Macromolecular Systems," *Acta Cryst.*, vol. 5, pp.1-6, 1952.

[33] G. Oster and D.P. Riley, "Scattering from Cylindrically Symmetric Systems," *Acta Cryst.*, vol. 5, pp.272-276, 1952.

[34] L. D. Cohen, R. D. Haracz, A. Cohen, and C. Acquista, "Scattering of light from arbitrarily oriented finite cylinders," *Appl. Opt.*, vol. 22, No. 5, pp. 742-748, 1983.

[35] L.E. Paramonov, "On optical equivalence of randomly oriented ellipsoidal and polydisperse spherical particles: The extinction, scattering and absorption cross-sections," *Opt. Spectrosc.*, vol.77, pp.660-663, 1994.

[36] M. Mi, J.W. Hovenier, C. Diminik, A. de Koter, and M.A. Yurkin, "Absorption and scattering properties of arbitrarily shaped particles in the Rayleigh domain: A rapid computational method and a theoretical foundation for statistical approach," *J. Quant. Spectrosc. Rad. Transf.*, vol. 97, pp.161-180, 2006.

[37] T. C. Grenfell and S. G.Warren, "Representation of a nonspherical ice particle by a collection of independent spheres for scattering and absorption of radiation," *J. Geophys Res.*, vol. 104, No. D24, pp. 31,697-31,709, 1999.

[38] S. P. Neshyba, T. C. Grenfell, and S. G. Warren, "Representation of a nonspherical ice particle by a collection of independent spheres for scattering and absorption of radiation: 2. Hexagonal columns and plates," *J. Geophys. Res.*, vol.108, No. D15, pp.4448, doi:10.1029/2002JD003302, 2003.

[39] T. C. Grenfell, S. P. Neshyba, and S. G. Warren, "Representation of a nonspherical ice particle by a collection of independent spheres for scattering and absorption of radiation: 3. Hollow columns and plates," *J. Geophys. Res.*, vol. 110, D17203, doi:10.1029/2005JD005811, 2005.

[40] I.E. Hansen, L.D. Travis, "Light scattering in planetary atmospheres," *Space Sci. Rev.*, vol.16, pp. 527-610, 1974.

[41] P. Latimer, "Light scattering by ellipsoids," *J. Colloid Interface Sci.*, vol. 53, pp. 102-109, 1975.

[42] F.D. Bryant and P. Latimer, "Optical efficiencies of large particles of arbitrary shape and orientation," *J. Colloid Interface Sci.*, vol. 30, pp.291-304, 1969.

[43] P. W. Barber and D. S. Wang, "Rayleigh-Gans-Debye applicability to scattering by nonspherical particles," *Appl. Opt.*, vol. 17, pp. 797-803, 1978.

[44] V.N. Lopatin, V.N. Aponasenko, V.S. Philimonov, K.A. Shapovalov, L.A. Shur, "Optical characteristics of "soft" biological particles suspensions and their connection with main factors that shape them," in *Static and Dynamic Light Scattering in Biology and Medicine*: SPI Proc. R. Nossal, R.Pecure, A.Priezzev, eds., vol. 1884, pp.356-364, 1993.

[45] N.V. Shepelevich, I.V. Prostakova, V.N. Lopatin, "Light-scattering by optically soft randomly oriented spheroids," *J. Quant. Spectrosc. Radiat. Transfer.*, vol. 70, pp. 375-381, 2001.

Printed by Books on Demand GmbH, Norderstedt / Germany